Finding Solutions for Environmental Conflicts

Environmental law is an increasingly important area of legal research. Given the increasingly interdependent web of global society and the significant steps being made towards environmental democracy in decision-making processes, there are few people that are untouched by environmental lawmaking processes.

At the same time, environmental law is at a crossroads. The command and control methodology that evolved in the 1960s and 1970s for air, land and water protection may have reached the limit of its environmental protection achievements. New life needs to be injected into our environmental protection regimes. This series seeks to press forward the boundaries of environmental law through innovative research into environmental protection standards, procedures, alternative instruments and case law. Adopting a wide interpretation of environmental law, it includes contributions from both leading and emerging European and international scholars.

Titles in the series include:

Whaling Diplomacy
Defining Issues in International Environmental Law
Alexander Gillespie

EU Climate Change Policy
The Challenge of New Regulatory Initiatives
Edited by Marjan Peeters and Kurt Deketelaere

Environmental Law in Development
Lessons from the Indonesian Experience
Edited by Michael Faure and Nicole Niessen

Finding Solutions for Environmental Conflicts
Power and Negotiation
Edward Christie

China and International Environmental Liability
Legal Remedies for Transboundary Pollution
Edited by Michael Faure and Song Ying

Finding Solutions for Environmental Conflicts

Power and Negotiation

Edward Christie
Barrister and Mediator, Queensland Bar, Australia

Formerly Associate Professor (Applied Ecology), Faculty of Environmental Sciences, Griffith University, Queensland, Australia

Formerly Environmental Member, Administrative Appeals Tribunal, Commonwealth of Australia

NEW HORIZONS IN ENVIRONMENTAL LAW

Edward Elgar
Cheltenham, UK • Northampton, MA, USA

Published by
Edward Elgar Publishing Limited
Glensanda House
Montpellier Parade
Cheltenham
Glos GL50 1UA
UK

Edward Elgar Publishing, Inc.
William Pratt House
9 Dewey Court
Northampton
Massachusetts 01060
USA

A catalogue record for this book
is available from the British Library

Library of Congress Control Number: 2008927695

ISBN 978 1 84720 070 9

Printed and bound in Great Britain by MPG Books Ltd, Bodmin, Cornwall

To the dancing stars in my life, Megan, Caleb,
Hannah and Ethan.

Contents

Foreword

There was a time when someone like Dr Edward ("Ted") Christie would have been described as a "renaissance man". That term was used to designate someone whose interests and accomplishments spanned a variety of academic and practical disciplines. In the 21st century, we need a new term, not only expressed in language that is gender-inclusive, but apt to cover the talents of a person who has the ability to combine the most up-to-date developments in several fields of thinking, so as to establish a new standard.

Whatever the new term might be, as this book demonstrates, it applies to Ted Christie. He was trained initially as an agricultural scientist. In the 1970s, when environmental concerns were becoming widely known, he was a pioneering environmental scientist. Later, he added law to his qualifications, and became a legal practitioner. Ted's career has combined teaching with distinction, practising as a barrister specialising in environmental disputes, and sitting as a member of a tribunal providing merits review of administrative decisions.

Once again, Ted's experience brought him to the forefront. He realised, earlier than most, that traditional methods of resolving legal disputes, based on conflict leading to the determination of a winner, were inadequate for many types of disputes, especially in the field of public law. He has been a leader in the field of alternative dispute resolution, particularly mediation.

Another area in which Ted has been ahead of the pack in dispute resolution, particularly in relation to environmental disputes, concerns the interests of Indigenous peoples. Realising that those interests may be affected vitally, even if they cannot be the subject of enforcement by law, Ted has been a passionate advocate for genuine consultation with Indigenous people about proposed developments on previously undeveloped lands.

It is not surprising that, with his breadth and depth of experience and expertise, Ted Christie has written a book that merits being described as revolutionary. As its title suggests, *Finding Solutions for Environmental Conflicts: Power and Negotiation* is a book about process in resolving disputes of particular kinds. It is more than merely that. It blends skilfully and entertainingly a number of themes.

At one level, this book is a legal textbook. It examines thoroughly a number of aspects of the way in which disputes about development proposals that will affect the environment are handled in three major common-law jurisdictions, the United States of America, the United Kingdom and Australia. This exercise in comparative law is useful in itself, because it assists policy choices when questions arise of the process to be adopted in relation to particular disputes, or disputes of particular kinds.

As a legal text, the book is also an excellent resource on the options available for alternative dispute resolution, and how to select from among those options. It enables the reader to understand fully the alternatives to traditional dispute resolution. It arms policy-makers with the information to choose more effective processes than legal proceedings that usually lead to unsatisfying settlements under the pressure of impending trials, or to judgments in which the court can deal only with the issues in the case, and can only rule in favour of one party or the other in relation to each of those issues

At another level, the book can be read as a scientific treatise. It provides valuable advice on the types of knowledge that are essential for making good decisions about development and the protection of the environment. It therefore promotes good outcomes, rather than outcomes based on "junk" science, supposition, rumour and ignorance.

Most of all, the book is a practical manual of how to produce optimal outcomes in environmental disputes. It offers advice about the steps that need to be taken to put in place procedures that are fair, rigorous and effective in producing those outcomes; procedures that leave no parties feeling that their points of view have not been considered, no viable option unexplored, and no room for lengthy and costly legal challenges to the outcomes.

If this book is read, and its contents are heeded, as widely as is justified, then the days of the application of traditional dispute-resolution procedures to environmental disputes should be over in the 21st century. Gone will be the days of steamrolling objectors, in order to push through development proposals at all costs; of withholding information unless forced to reveal it after costly legal proceedings; of announcing that, of course there will be "consultation", but the project will go ahead in any event; of slanted or superficial environmental impact statements, designed to downplay the very real problems that will arise when it is too late to prevent them; of footing massive legal bills to defend at all costs the outcomes of processes that have denied procedural fairness to those participating in them; and of winner-take-all judgments that end up satisfying nobody.

Perhaps it is excessively utopian to think that these things can happen, but this book offers a very real prospect that they can. It will be a compulsory item on the bookshelves of every community organisation aiming at the protection of the environment from excessive destruction by development. It will never gather dust on those bookshelves. Bureaucrats and those whose task is the resolution of environmental disputes will ignore it at their peril; they will realise that it offers advice that will save time and money in processes that are currently too long and costly. Lawyers who act for parties to environmental disputes will find it invaluable as a source of authorities on the law, and of practical guidance. Developers will find that they need to read it before making applications for approval of projects, lest they make those applications without understanding what will be required of them, if they are to succeed.

To describe this book as revolutionary is not to resort to hyperbole. It will bring about significant change. Ted Christie, renaissance man (or its 21st century equivalent), deserves acclamation for having written it.

Peter R.A. Gray
Judges' Chambers
Federal Court of Australia
Melbourne
January 2008

Acknowledgements

The cross-disciplinary collaboration and use of knowledge to resolve environmental conflicts, the central feature of this book, has been shaped by a number of people to whom I owe a debt of gratitude. My transition from environmental science to environmental law was triggered by Dr Len Webb, an Australian rainforest ecologist of international renown. Following a bewildering experience he had as an expert witness in court in the rainforest logging and conservation conflicts in Australia in the late 1970s, Len Webb persuaded me to take a directional career shift from environmental science and academe to law and practice because of his steadfast belief that the time had come for science to establish a more effective control in resolving environmental conflicts. Alternative dispute resolution was not an option at this time in Australia.

Following admission to legal practice, I was privileged to work with the Commission Chairman Tony Fitzgerald QC, a distinguished lawyer and jurist, in a public interest environmental inquiry in 1990–91 (the *Fraser Island Inquiry*) and to observe his professional and interpersonal skills first hand. The manner in which Tony Fitzgerald conducted the inquiry recognised the paramount need for the parties to have trust in the conflict resolution process as well as trust in the dispute resolver — intangible values that were a key feature of this inquiry. Moreover, it recognised that the process for resolving information conflicts could be effectively undertaken independently of the litigation model and its associated procedures for the examination of evidence.

Immediately after this inquiry, I was appointed to the Commonwealth of Australia's Administrative Appeals Tribunal, in 1992, for a merits review of a multi-party, factually complex environmental conflict related to the future use of the Great Barrier Reef Marine Park. The Tribunal was presided over by an eminent judge of the Federal Court of Australia, Justice Peter Gray. The case involved the examination of the evidence of 69 expert witnesses, over 12,000 pages of transcript and 22 weeks of hearing. Each day, I saw probably the most just and effective demonstration of the principles of procedural fairness (or "procedural due process" in the United States) being applied by Justice Gray to provide all parties with an equal opportunity to shape the final decision. All of these observations left an indelible impression on me and later became the foundation for the approach I have taken for developing and applying processes to assess, to manage and to resolve environmental conflicts.

Transforming real problems from professional practice into a book that could effectively communicate cross-disciplinary collaboration and use of knowledge was optimised through convening and teaching cross-disciplinary courses at Griffith University, from 1996, on a part-time basis: Environmental Law to Faculty of Law undergraduates and ADR to Master of Environmental Management post-graduates. These students were both committed and stimulating to teach. I owe them all a special thanks, as interacting with them over time enabled me to crystallise my professional expertise on the environment into a cross-disciplinary book on problem-solving.

Given that this book is based on cross-disciplinary collaboration and use of knowledge, for those who require a more detailed coverage of environmental law, administrative law or ADR and negotiation, the following excellent texts and materials, referred to in this book, should be reviewed: *Environmental Law (Australia)* — Bates (2006); *Environmental Law (UK)* — Bell and McGillivray (2006) and Stookes (2005); *Environmental Law (USA)* — Plater *et al.* (2004), Ferrey (2004) and, especially, the publications of the United States Environmental Law Institute; *Administrative Law (Australia)* — Halsbury's Laws of Australia; *Administrative Law (UK)* — Wade and Forsyth (2004); *Administrative Law (USA)* — Asimow *et al.* (1998); *Negotiation* — Fisher *et al.* (1991) and Susskind *et al.* (2000); *ADR* — Sourdin (2005).

In terms of the preparation and production of the book, it is a pleasure to acknowledge the unsurpassed research assistance and support provided to me by Mary Novello over time and, above all, Denise Burton, for her untiring and prolonged efforts in persevering to transform my hieroglyphics into a word processed manuscript. Special thanks are owed to Bev Jeppessen who prepared the camera ready copy with consummate professionalism.

It is also a pleasure to acknowledge the kind support I received over time from my publisher, Edward Elgar, most particularly Commissioning Editors, Catherine Elgar and Felicity Plester, Senior Editor, Alexandra O'Connell and Deputy Managing Editor, Karen McCarthy. I am especially grateful to Catherine Elgar, who was instrumental, during our first meeting in Australia, in recommending that my proposal for a cross-disciplinary book on resolving environmental conflicts should extend the legal principles more broadly from an Australian to a global context.

Finally, I would like to express special thanks to my family, especially my wife Meg, Tim and Melody, for inspiring and encouraging me to follow and to believe in the changing directions of the professional pathways that I have pursued over time. Tolerance and patience were other virtues Meg possessed which enabled me to embark on a journey interspersed with chance. For me, it is a great pleasure to be able to dedicate this book to my wife Meg.

1. Introduction

"A land ethic then reflects the existence of an ecological conscience, and this in turn reflects a conviction of individual responsibility for the health of the land. Health is the capacity for the land for self-renewal. Conservation is our effort to understand and preserve this capacity." **Aldo Leopold (1949)**

"Our story is in the land, It is written in these sacred places.
My children will look after these places. That's the law."
Big Bill Neidjie, Bunti Clan, Gagadju Language Group (1982)
(Cited by Mosley, 1988)

"[Sustainability problems] cannot be resolved without shared responsibility and joint action — without the engagement and the assumption of responsibility by the stakeholders and the citizens in the communities where these problems are manifesting themselves."
John Kitzhaber (2004) (Former Governor of Oregon, USA)

" ... in 1981, alternative dispute resolution was widely regarded as an unconventional, inferior substitute for the only proper means of resolving disputes: namely, litigation in publicly funded courts and tribunals. That attitude has changed dramatically ... and it seems only a matter of time before Australian governments follow the lead of the UK Government which, in 2001, pledged that all its departments and agencies would settle all legal disputes by ADR wherever possible."
Tony Fitzgerald QC (2006) (Inaugural President, Queensland Court of Appeal 1991–1998)

Conflicts over environmental problems such as biodiversity, biotechnology, environmental impact assessment, hazardous chemicals, Indigenous land interests and the "greenhouse effect" are not necessarily legalistic problems but land use problems. What many environmental conflicts have in common is that they are all global problems in sustainability – involving ecological, economic, social and cultural considerations. In addition, environmental conflicts will almost certainly be associated with scientific uncertainty in the information that is available for decision-making, so resulting in information being a primary source of conflict. In circumstances where inadequate information and uncertainty exists, litigation results in a final adjudication of fact. The party which has the onus of proof must persuade the court according to the requisite standard of proof or fail. In contrast, science would defer a decision in these circumstances until more research and conclusive evidence were available. The finality which is crucial for the legal method is not present in the scientific method as it operates under no deadlines for time (Martin, 1977).

Environmental problems have sometimes been described by analogy to water continually flowing out of a shower faucet. Control resides at the shower tap. The traditional role of science for resolving environmental problems has been to mop up the puddles on the shower floor by undertaking research into the nature of the problem — but with little direct control in conflict resolution. Lawyers, in contrast to scientists, have had almost complete control for resolving environmental conflicts because of the reliance on litigation as the pathway for adjudicating environmental conflicts. The central issue is not one of control. Rather, how can environmental conflicts be most effectively resolved? Through litigation, where a court makes a determination and imposes its decision on the parties — or alternatively, by the parties themselves, finding their own mutually agreed solution by negotiation?

THE GROWTH OF ENVIRONMENTAL LAW

Whilst legislation directed at environmental matters has been introduced throughout the world for over a century, environmental law, as a discrete area of professional legal practice, is a relatively new contemporary direction. It has added a new dimension to environmental decision-making. The person first given credit for using the term "environmental law" was US Attorney Paul Yannacone, a lawyer who was to become a central figure whilst representing Vietnam veterans in the "Agent Orange" litigation in the United States in the 1980s. Whilst arguing in a New York court in the 1960s for a ban on DDT spraying to safeguard public health, Yannacone was asked by the trial judge as to the legal principles which could possibly justify a ban. In his reply Yannacone coined the, then, new phrase — "Environmental Law" (Schuck, 1986).

Some 40 years later, with environmental litigation now a well-accepted part of environmental regulatory control throughout the world, a further significant change in professional legal practice that has implications for conflict management and resolution has been the enormous growth in environmental legislation. There has also been an upsurge in the number of international environmental conventions introduced which reflects the situation that many environmental problems are no longer simply national problems but require a global approach for their resolution. International conventions represent a supplementary source of potential legal rights at the national level. There is little dispute that contemporary environmental legislation has become both technical and complex. The growth in environmental legislation has resulted in new rights and duties arising under legislation and increased statutory powers available to Government agencies in their administrative decision-making functions — as well as increased litigation.

Further obstacles for managing and resolving environmental conflicts, associated with the growth of environmental legislation, relate to statutory interpretation. This is particularly the case when environmental legislation incorporates key scientific terms and concepts, but fails to define their meaning consistently with their accepted scientific usage; or alternatively, where a statute fails to provide any definition (Christie, 1990). For example, in 1989 planning legislation for the State of Queensland, Australia, did not provide any meaning for the term "environment" — yet the statute required the planning authority to consider "*any deleterious effect on the environment*" arising from a proposed development. In the case *Murphy & Another v The Crown* (1989) 68 LGRA 286, the issue was whether a proposed subdivision of land adjacent to a beach — the site of a world renowned turtle rookery, should be permitted in the interests of the turtle population. In a 2:1 decision, the Full Court of the Supreme Court (the ultimate or Appeal Court of the State), in its construction of the term "environment", concluded that living organisms should not be included in the definition of the environment. However, the legislation was subsequently amended to provide a definition for the term environment which included living organisms. Consequently, poorly drafted environmental legislation can only create difficulty for finding solutions for environmental conflicts, especially when legal rights are dependent on the meaning of scientific terms and concepts contained in environmental legislation. In the extreme case, the environmental regulatory control purpose of the legislation may prove to be an illusory bargain.

There are also a number of important characteristics which today's practitioners of environmental law clearly recognise. Because environmental conflicts are significantly founded on scientific and technical issues, the law turns to science for answers to factual questions that may be beyond the understanding and knowledge of non-scientists. Lawyers, who may not have had any scientific training, are then required to understand the scientific evidence, review it for its reliability and to then apply that evidence to the conflict in question. Significantly, the time involved in preparing evidence for environmental litigation and, in turn, the length of the hearing and the examination of evidence by a court, will be inordinately weighted towards scientific evidence rather than to questions of law.

GLOBAL TRENDS IN THE RESOLUTION OF ENVIRONMENTAL CONFLICTS

There has been a significant change in terms of the range of available processes for resolving conflicts from the time that the term "environmental law" was first used by Paul Yannacone and where the litigation pathway prevailed. Alternative dispute resolution ("ADR") and negotiation has now moved from its past perception, as a quirky process for conflict resolution, to its increased acceptance and use in many areas of professional legal practice — including its implementation in legislation and policy, e.g. for the United States, the *Alternative Dispute Resolution Act of 1998* and the *Memorandum on Environmental Conflict Resolution (2005)*. Negotiation is the primary form of dispute resolution and therefore is the foundation for a wide range of ADR processes, such as mediation, independent expert appraisal and neutral evaluation — processes that are described as "structured assisted negotiation dispute resolution processes".

ADR and litigation are two complementary, but entirely different, approaches for managing and resolving environmental conflict. However, ADR is not a universal panacea that can necessarily be applied to resolve all environmental conflicts. There will be some environmental problems where litigation is the only feasible pathway to pursue. For contemporary environmental problems, the central issue is to be able to assess the circumstances when either ADR or litigation is the most appropriate process for managing and resolving a specific conflict — or a stage of the conflict.

Consequently, finding solutions for environmental conflicts can take a number of forms. The most common pathway is to initiate a court action. Following an action being filed, one approach to resolve the matter used by lawyers acting for the parties to a dispute is to seek an out of court (or pre-hearing) settlement before the hearing commences. Where an out of court settlement does not result in an agreed outcome, ADR is often the next step towards conflict resolution, particularly where ADR is prescribed as a court-annexed process. In this situation, the parties to a dispute may have to undertake ADR either as a compulsory or voluntary requirement, as prescribed by legislation. In contrast to an out of court settlement, the parties will be present with their lawyers as ADR and negotiations proceed. Where the parties fail to reach any solution, either through out of court settlement or a negotiated agreement through the court's annexed ADR processes, the matter may then go on to a hearing. It then becomes the court's role to determine the matter and to impose its decision on the parties.

Community consultation and engagement processes, public participation processes and public inquiries have been the main way Governments have provided opportunities for parties to be involved in public interest environmental conflicts outside the courts and litigation. Compared with resolving environmental conflicts through ADR, these processes limit the power the parties have in the extent of shared involvement in fact-finding as well as for shared responsibility in the decision that is ultimately made. This limitation may be offset when ADR — external and independent to the court-annexed ADR pathway — is used to resolve public interest environmental issues involving site-specific and policy-level conflicts. The benefit ADR and negotiation brings to environmental conflicts is to provide the parties with opportunities to engage the Government agency through constructive collaborative problem-solving to manage and to resolve an environmental conflict.

Many parties engaged in an environmental conflict continue to rely on litigation as the preferred model for resolving environmental conflicts. ADR processes, such as mediation, have sometimes been described as inadequate when applied to environmental conflicts (Horn, 2005). However, such an observation fails to recognise the distinguishing characteristics of environmental conflict resolution. It assumes that only one process will be required to manage and resolve a specific conflict. Conflicts associated with the environment and natural resources frequently involve multiple, recurring subjects of tension and conflict. Moreover, it may well be the situation that no single means for managing and resolving complex, interrelated and persistent sources of conflict will necessarily be effective. Instead, the parties may find that they need to pursue several different processes in order to manage the conflict and to reach consensual agreement. Accordingly, a key issue for resolving environmental conflicts is to know when to rely on a specific conflict resolution process, and when this process is seen as being appropriate for managing or resolving a specific conflict, or a stage of the conflict (Arnold and Jewell, 2002).

LITIGATION OR NEGOTIATION: BATNA ANALYSIS

There is an important distinction between litigation and negotiation in terms of the "settling of disputes" and the "resolution of conflict" that needs to be recognised, particularly if the goal is to find solutions for an environmental conflict. Litigation through courts settles disputes. As Street (2005) points out, "*adjudicative or determinative processes are not dispute **resolution*** *processes*". That is, judges do not *resolve* conflicts that are litigated in courts. Rather, judges *adjudicate* on disputes and then impose a binding decision on the parties. In the process of adjudication, judges *settle* questions of law in order to decide the ultimate issue. Furthermore, as Kirby (1992) points out, settlement of a particular dispute results in the underlying causes of conflict being unresolved and remaining as a source of resentment or irritation. Consequently, the conflict may re-emerge at some later date and require resolution.

In contrast, the goal of ADR and negotiation is conflict resolution and so can be distinguished from litigation. For most ADR processes (but arbitration is one exception[1]), the dispute resolver's role is to assist the parties to resolve their conflict by finding their own solution through a negotiated agreement. Moreover, the dispute resolver, unlike a judge, does not decide the issues or impose any decision on the parties that would be binding on them.[2] In contrast to litigation, ADR enables a solution for the immediate conflict to be negotiated — and may also incorporate outcomes for potential environmental problems should they emerge at some future date.

The decision to litigate or to negotiate can be evaluated by each side comparing the pattern of strengths and weaknesses of their case relative to the other side. If litigation is the preferred pathway for conflict resolution, the potential risk of losing the action needs to be assessed. Alternatively, if negotiation is the preferred pathway for conflict resolution, some assessment of the sources of power needs to be undertaken as the outcome of negotiation will reflect the power available to the parties. A strategic means for deciding whether to litigate or to negotiate is to consider a key element of principled negotiation advanced by Fisher *et al.* (1991) — the best alternative to a negotiated agreement ("BATNA"), i.e. the course of action to be taken if no agreement can be reached by negotiation. BATNA analysis is seen as a significant source of power for screening the appropriateness of litigation or negotiation — as well as consultation, participation and political lobbying — for resolving an environmental conflict. A BATNA analysis enables each party to consider what may happen to them, in terms of an outcome, if a negotiated agreement cannot be reached.

---

[1] Arbitration was one of the original forms of ADR introduced into western countries following general dissatisfaction with courts. However, arbitration became criticised as it was associated with many of the problems of adversarial litigation which it was intended to replace.

[2] Arbitration may be an exception to other ADR processes, in this regard.

ENVIRONMENTAL NEGOTIATIONS AND KNOWLEDGE POWER

Any outcome negotiated by consensus will reflect the power relationships between the parties themselves. Power is a key concept central to conflict resolution. In terms of understanding conflict dynamics, Mayer (2000) defines power as "the ability to get one's needs met and to further one's goals". Gerald Cormick, who mediated one of the earliest environmental conflicts in the United States — a proposed dam on the floodplain on the Snoqualmie River in Washington State in 1973 — argued that mediation was only suitable "*when there is a relative balance of power between the parties*" (Melling, 1995). However, Mayer (2000) states that it is more meaningful to consider differences in power, the sources of power and vulnerability to the power of the other side, rather than whether power can be balanced. That is, it is more relevant to consider whether the parties to a dispute have "*an adequate basis of power to participate effectively in the conflict*".

Information conflicts predominate in any environmental conflict and differences will almost certainly exist in the information available to each party. Consequently, to participate effectively in an environmental conflict, parties must be aware of, as well as understand, the relevant sources of information. Knowledge of legal rights together with knowledge and understanding of scientific data and information should be recognised by Government agencies, proponents of development projects and the community as significant sources of knowledge power. For environmental conflicts, knowledge power of legal rights and scientific data and information have two complementary elements that are central to the concept of power relationships: sharing of power and effective participation. Both elements are central to reaching consensual agreement through negotiation. Knowledge power exercised in this manner can also offset the concerns held by a party on their vulnerability to the power of the other side.

(a) Sharing Power and Legal Rights

Where power arising from knowledge of legal rights is properly exercised, opportunities are provided for Government, proponents of development projects and the community to share power to negotiate solutions. Power, effectively exercised in this regard, should be seen as providing a "trigger" to share power to manage an environmental conflict by joint fact-finding or to resolve an environmental conflict by a joint problem-solving approach to reach a negotiated agreement.

(b) Power, Effective Participation and Science

The application of knowledge power arising from an understanding of the scientific data and information for environmental management and protection also facilitates creative outcomes being developed and evaluated by the parties. Any inequality in bargaining position associated with data conflicts may be offset where a collaborative fact-finding process is based on a common database agreed to be the relevant and reliable scientific evidence that is available and which is understood by all parties.

FINDING SOLUTIONS FOR ENVIRONMENTAL CONFLICTS: THE CHALLENGE

Given the increased global awareness and acceptance of ADR and negotiation as an appropriate process for resolving conflicts, generally, it should also be recognised that the management and resolution of environmental conflicts through ADR requires effective integration between law and science. Finding solutions for environmental conflicts through ADR should not be seen as the exclusive domain of law, or as the sole province of science. Consequently, the scope of this book is to focus on a cross-disciplinary, problem-solving approach that integrates law and science within the framework provided by the principles and processes for ADR and negotiation.

(a) Cross-Disciplinary Use of Knowledge

Any failure to recognise the cross-disciplinary nature of environmental law can only contribute to concerns whether satisfactory environmental outcomes can be achieved through ADR and negotiation. Negotiated agreements arising through ADR should not be open to claims — as has sometimes been the case in the past for litigation — of the poor quality scientific fabric used as the basis for legal decisions (Burger, 1988).

Where legal rights are in issue, the boundaries for the scientific evidence and information are set by questions of law, as prescribed by the relevant statute. For real progress to be made in integrating law and science to manage and resolve environmental conflicts, the scientific evidence and information must conform to the standards and criteria of science. In addition, where scientific terms and concepts are prescribed as questions of law in legislation, they must have a statutory meaning that is consistent with their scientific usage.

(b) Dispute Resolver Skills

The need for a cross-disciplinary problem-solving approach for managing and resolving environmental conflicts also adds a further element to the challenge for finding solutions for environmental conflicts. In order to optimise the use of ADR and negotiation for managing and resolving environmental conflicts, it needs to be recognised that today's alternative "dispute resolver" requires both ADR process skills, as well as knowledge of the substantive content or subject matter of the conflict, that is, law and/or science. This is a significant departure from the general approach for ADR, where process skills have frequently been considered to be the only criterion necessary for the professional practice as an alternative dispute resolver. The need for the alternative dispute resolver to have both process skills and subject matter skills for managing and resolving environmental conflicts can be best understood when it is recognised that any environmental conflict will be characterised by the diversity of factual evidence and information arising from sources such as the chemical, ecological, economic, social and statistical sciences as well as anthropology. Moreover, the scientific facts will not only be complex, but also numerous. Consequently, the use of ADR along the sequential stages of conflict assessment, management and resolution, for environmental conflicts, may be facilitated by a lawyer as the alternative dispute resolver with process and subject matter skills. Additionally, an environmental non-legal professional would have a similar role, along these sequential stages, but only where factual issues were involved, i.e. where the interpretation and application of questions of law were not in issue.

(c) The Community

Bonine (1998) considers that the worldwide move towards citizen enforcement of environmental laws and obligations is one of the most striking contributions that environmental law has made to civilisation worldwide at the end of the 20th century. Critics of the use of negotiation often point to constraints imposed by the non-compromising stance taken by environmental groups to resolve environmental conflicts. For example, the observation made by Kelty (1991) that a view projected by some environmental groups was that "*we are not negotiators, we are the purveyors of ideas, we do not have to compromise, we do not have to reach agreement with you, we can oppose you to the end*" may appear as a major obstacle for pursuing negotiated agreements. However, one contemporary North American analysis (Bingham and Haygood, 1986) of 132 environmental conflicts involving land use, public lands, water resources, energy, air quality and toxic chemicals suggests this perception is not necessarily a true representation for all environmental groups. Their study, for a ten year period, provides insight into the potential to find solutions for environmental conflicts by negotiation where a varied and broad mix of parties was involved — government, industry, local citizen groups and environmental bodies. Overall, agreement was reached in 78 per cent of cases, with little difference between siting disputes and policy disagreements. There was also a high correlation between the resolution of the disputed issues and the subsequent implementation of the agreement for siting disputes: 80 per cent being fully implemented and 13 per cent partially implemented (Bingham and Haygood, 1986). These findings offset criticism created by any perception that environmental groups will always take a non-compromising stance, regardless of the nature of the conflict and so provides a strong basis for justifying the use of negotiation to find solutions for environmental conflicts.

RESOLVING ENVIRONMENTAL CONFLICTS BY NEGOTIATION: LAW, SCIENCE AND ADR IN PERSPECTIVE

The aim of this book is to find solutions for site-specific and policy-level environmental conflicts through ADR and negotiation. The approach in the book is not to simply focus on law or science or ADR as discrete sources of knowledge. Rather, the book links (i) legal rights and principles in environmental and administrative law to (ii) accepted scientific concepts and information needs for environmental management and protection with (iii) established principles and processes for ADR and negotiation. There are a number of significant distinguishing features for this book:

- First, the book is cross-disciplinary as it integrates law, science and ADR so as to achieve the goal of finding solutions for managing and resolving environmental conflicts. The integration across disciplines is a major feature of this book as it has its foundation in the understanding and application of cross-disciplinary knowledge.
- The book provides an understanding of key legal rights and principles contained in environmental and administrative law for Australia, the UK/EU and the United States, together with relevant international environmental conventions. The legal rights reviewed are especially significant as sources of power for resolving environmental conflicts.

- Clearly, this book cannot cover all environmental legislation or scientific concepts for environmental management and protection or environmental problems that confront society. Rather, the approach taken throughout is to focus on key unifying principles in law, science and ADR and negotiation that have wide application for conflict resolution. The goal of finding solutions for environmental conflicts, that are contemporary sources of global concern, extends to *potential* conflicts that pose risks to the environment and for human health.
- Collaborative approaches to fact-finding, problem-solving and decision-making, specific for the characteristics of environmental conflicts, are achieved through the application of the established concepts for principled negotiation (Fisher *et al.*, 1991) and the mutual gains approach to negotiation (Susskind *et al.*, 2000).
- It has long been recognised that a key advantage of ADR and negotiation is that it has the potential to provide a greater range of solutions and outcomes for conflict resolution, relative to litigation. This advantage is particularly significant when sustainability and the environment is in issue. However, in order to optimise this advantage inherent in the ADR process, a significant element of principled negotiation (Fisher *et al.*, 1991) must be effectively applied: the parties to an environmental conflict must be able to develop a range of "*creative options for mutual gain*". An understanding of the accepted scientific concepts for environmental management and protection forms the basis for being able to develop creative solutions for mutual gain.
- The book develops and extends the well-established conflict resolution concept of knowledge power as a basis for using ADR to find solutions for environmental conflicts by negotiation. Knowledge power is considered, not only in the context of legal rights but also from the perspective of environmental science. Given that scientific information (or data) conflicts will invariably be a significant source of conflict for environmental problems, the book considers knowledge power in the context of a further element of principled negotiation (Fisher *et al.*, 1991): specifically, "*the use of objective criteria*" to understand and to evaluate scientific information. This element is the foundation for environmental problem-solving.
- Indigenous traditional knowledge represents a different perspective on the local environment, causes and consequences of environmental change, natural resource management and social interactions (Abaza *et al.*, 2004). Where Indigenous communities are provided with legal rights in environmental legislation, such as recognising the role of Indigenous people to promote the use of traditional knowledge in conservation and sustainability matters, law represents a significant source of knowledge

power to facilitate the use of the collaborative problem-solving procedures of ADR and negotiation. In addition, where sustainability is prescribed as a legal right in environmental legislation, Indigenous people would also be given an opportunity to pursue positive social and economic considerations as part of any negotiated agreement associated with a major development proposal. The challenge is for developers to be coopted through the negotiation process and drawn away from using more traditional methods of dealing with Indigenous people — often involving ignoring them or mobilising the Government to protect developer interests (O'Faircheallaigh, 1999).

- Environmental conflicts will invariably have multi-party involvement. Steiner (2006) states that the task for finding solutions for biodiversity conflicts, today, must be addressed by new alliances across all sectors of society that share an interest in conservation action for biodiversity — and not only by environmentalists. Whilst this statement was made in the context of biodiversity conflicts, the potential for forming new alliances has broader application for resolving environmental conflicts by negotiation. An alliance (or a "coalition") can be formed for an environmental conflict where the parties constituting an alliance have common interests, goals or aspirations in relation to the conflict. Alternatively, an alliance can be formed where the parties are likely to be affected in a similar manner, by an administrative decision to be made by a Government agency for a proposed action having potential adverse consequences for the environment.
- Finding solutions for environmental conflicts requires different ADR processes for the sequential stages of conflict assessment, conflict management and conflict resolution. The approach to cross-disciplinary collaboration links the appropriate skills required by a dispute resolver — lawyer or environmental professional — to the specific ADR process used for each stage of the conflict.
- Finally, an additional feature of the book will be to address the "*gatekeeper role*" for the lawyer, the environmental professional and the community as part of a BATNA analysis. Specifically, consideration of objective criteria that can be used to decide whether to use either negotiation or litigation for managing or resolving a specific environmental conflict.

ORGANISATION OF THE BOOK

The book is structured into three parts. The first part (Chapters 2–4), reviews knowledge power in terms of unifying legal rights and principles in environmental and administrative law. Understanding these legal rights and principles has application to power-sharing to facilitate the use of negotiation to resolve environmental conflicts. This part also considers the various elements of any environmental conflict and distinguishes their role for decision-making in litigation and negotiation. In addition, a critique of participatory approaches is given in order to distinguish significant differences between community consultation, public participation and ADR for public interest environmental conflict resolution.

The second part (Chapters 5–9) reviews knowledge power in terms of scientific principles and concepts for environmental management and protection that have applications for effective participation in joint problem-solving when negotiated agreements are sought. An analysis of objective information needs for environmental decision-making, including ADR processes to address information conflicts, is provided for a number of contemporary global environmental problems: sustainability, biodiversity, EIA, biotechnology and hazardous chemicals.

The final part (Chapter 10) integrates established principles from ADR and negotiation with cross-disciplinary knowledge from law and science into a collaborative model for problem-solving of environmental conflicts by negotiation.

SUMMARY

The book has been written to meet the requirements of any environmental professional — lawyer, scientist, engineer, planner... — who, directly or indirectly, may be involved in planning or development conflicts when the environment is in issue, particularly where the aim is to find solutions that are politically and legally realisable. For the lawyer, the book, with its focus on understanding and integrating legal principles and scientific concepts, consolidates opportunities for finding solutions for environmental conflicts by negotiation, through the use of ADR processes appropriate for conflict assessment, management and resolution.

For the non-lawyer environmental professional, the book provides opportunities for the use of ADR, primarily to manage environmental conflicts. In addition, opportunities are identified for resolving environmental conflicts, by ADR and negotiation but in quite specific situations, i.e. when the interpretation and application of questions of law are not in issue and only factual (scientific) issues are in dispute.

Moreover, given the heightened professional knowledge of environmental issues within the community today, the book provides an understanding of the potential pathways for private citizens, the Indigenous community, the general community and environmental groups who are seeking more direct and effective inputs into resolving environmental conflicts through ADR and negotiation.

2. Principles and concepts in environmental decision-making

"Engineers [and scientists] are a pain to deal with ... [in] an environmental negotiation. They speak a language that 99% of the human race cannot understand. They have two hemispheres in their brains, just like the rest of us, but they insist on using only one of them, the logical analytical side."

Gorczynski (1991)

"I understand that the Act [the *Environmental Protection Act 1994 (Qld)*] is an example of an Act written in 'plain English'. My reading of the Act for the purpose of this case convinces me the language used is not plain and whether it is entirely in English could be debated."

Skoien SJ in *Richards & Sons Pty Ltd v Ipswich City Council* (1998) QPELR 303

"The Appellant has attempted to distinguish the factual situation in this case from that in Renfroe v Higgins Rack Coating and Manufacturing Co Inc. He didn't. We couldn't. Affirmed. Costs to appellee. All concurred."

Decision of **Gillis J** in *Denny v Radar Industries Inc.* (1970) 28 Mich App 294
(Cited by Sheller J, 1999)

"... the [adversarial] system works best in the simpler cases ... Where the facts became very complex and numerous, the [adversarial] system worked less well."

Pincus J (1988) (Federal Court of Australia Justice, as he then was)

"The facts produced in an environmental mediation are liable to be incomplete and inaccurate."

Plater *et al.* (2004)

INTRODUCTION

Environmental conflicts are characterised by the need for cross-disciplinary information from many sources of science, anthropology and law through to Government policy. The components shown in Figure 2.1 do not necessarily always occur in every environmental conflict. A relatively simple conflict, such as an application for a licence for an environmental activity, will impact on a number of these components. A complex public interest environmental conflict, such as a major development proposal, will require information from a very wide range of professional disciplines. In addition, a feature of many environmental conflicts is that there may have been only limited research directly undertaken on the disputed issues, so that some degree of scientific uncertainty will prevail. In this situation, the information available for environmental decision-making may have to be derived indirectly through the scientific method of inference, or through expert scientific opinion, rather than facts acquired directly through experimentation.

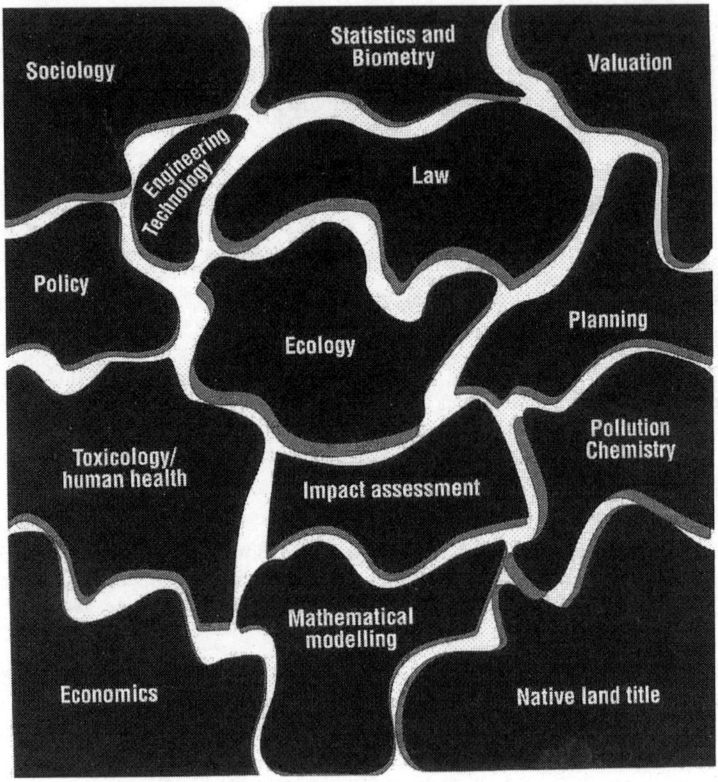

Figure 2.1 Information needs for resolving environmental conflicts

In order for an environmental problem to be effectively resolved the potential sources of conflict that may have triggered off the problem need to be identified. The sources of conflict are central for addressing the question of power and effective participation and, in turn, the choice of process for resolving conflict. Moore (1986) has summarised the various sources that contribute to the causation of any conflict:

i. Conflicts over **information** (or **data**) arising through a lack of information, misinformation, scientific uncertainty, different interpretations of the same information or different opinions as to what information is relevant. Adoption of experimental and research methods, where there are no standard protocols, exacerbate information conflicts;
ii. Conflicts over **values** that occur because of different cultures, ways of life, ideologies and ethics. The use of different criteria to evaluate ideas or behaviour confounds value conflicts;
iii. Conflicts over **interests**. Interests refer to the needs, concerns or hopes that a party seeks to have satisfied through conflict resolution. Concerns over potential adverse environmental consequences from an action or activity would be a substantive interest;
iv. **Structural** conflicts arise where there is an imbalance of power, for example, because of differences in financial and human resources or knowledge of the substantive issues; and
v. Conflicts over **relationships** caused when strong emotions become entangled with the objective facts of the problem. Persistent negative behaviour of a party aggravates the situation.

The first three sources of conflict relate to the subject matter of the conflict, whereas the latter two sources are associated with the characteristics of the parties to a dispute. Regardless of the type of problem that parties seek to resolve, the sources that contribute to causation of the conflict, as identified by Moore, will all be present. However, the balance between each source and their relative significance will vary depending on the nature of the conflict. Where the environment is in issue, conflicts over information and values will invariably be the primary sources of conflict. Carpenter and Kennedy (1985) point to the need to recognise that limitations in the available information will be a source of conflict for mediating environmental conflicts, particularly complex environmental problems. Smith (1987) has observed that misinformation is a feature of environmental conflicts and refers to mediation being relevant to the 60 to 70 per cent of environmental conflicts that arise by misinformation.

An analysis of Figure 2.1 provides a broader categorisation of information as a source of conflict for an environmental problem, other than in terms of the scientific discipline. The information components can be subdivided into the following elements (or "categories"), which may all have a role in shaping the outcome of the environmental decision-making process:

i. questions of law;
ii. international environmental treaties and conventions;
iii. facts (expert scientific opinion or lay evidence);
iv. policy; and
v. values.

The weight that is given to each element for the ultimate outcome is quite different depending on whether litigation or negotiation is used. Clearly, this is a key factor that should be central to any conflict assessment. Consequently, an understanding of each element for environmental decision-making is essential in deciding whether to proceed with either litigation, negotiation or other public participatory approaches such as community consultation or political lobbying.

QUESTIONS OF LAW

Understanding the meaning of the legal concept of a "question of law" frequently causes some difficulty for non-lawyers. In its simplest sense, a question of law represents a matter that relates to the application or interpretation of the law. Legal rights prescribed in a statute are the source of questions of law. Legal rights prescribed in environmental legislation often incorporate scientific terms and concepts which then become questions of law that must be determined through a review of the scientific evidence.

Questions of law are very significant in that they set the boundaries for the scientific evidence which must be addressed by experts for each party. The body of law arising from judicial decisions that interpret questions of law and legal rights contained in legislation is generally referred to as the common law. The common law is a significant source of legal principles to complement rights and obligations contained in environmental legislation; together, they represent a significant source of knowledge power for evaluating whether legal rights provide an adequate basis for effectively participating in negotiating agreements. Legal principles developed as part of the common law as precedent in each country is only binding on courts in that country; additionally, a lower court is bound to follow the precedent of a court higher in the hierarchy, in the same jurisdiction. Legal principles developed as part of the common law in one country may have persuasive value in other countries, but they are not binding.

A court will only give effect to the meaning for the specific scientific term or concept as it is defined in the statute itself, rather than a definition contained in a scientific publication. However, where a scientific term or concept is prescribed in a statute, but is not defined, the court will rely on the expert evidence at the court hearing to assist the court to arrive at its own construction, or interpretation, for the meaning of the scientific term or concept. A legislative example relating to the interpretation of a question of law that incorporates a scientific term or concept is the term "critical habitat". Critical habitat is a concept that is found in global legislation for the regulatory control of threatened species. A question of law may be whether the statutory meaning for critical habitat for a threatened species of fauna has been correctly interpreted and applied, according to its meaning as defined in the relevant statute. In its consideration of the interpretation of a question of law, a court may review legal principles that provide precedent from cases involving similar fact situations for the specific statute and the associated question of law.

Questions of law are matters for adjudication within the exclusive domain of the court and so are reserved for the court to decide. Moreover, it should be understood that although an issue may turn on a question of fact, a question of law is the ultimate issue for the court to decide and is determined by the judge applying the facts to the question of law. That is, the decision end-point for litigation is reached when the findings on fact made by a judge are applied to the relevant question(s) of law that have been prescribed by the statute.

National (Domestic) Sources of Environmental Legislation

Legislation refers to a statute or regulations made in accordance with a statute. In Australia, the source of environmental legislation that is enforceable by action of the Federal Government is the Federal (or "Commonwealth") parliament. The parliaments of the States and Territories of Australia implement statutes that primarily apply within the boundaries of the enacting State or Territory. Similarly, in the United States, the Constitution provides Congress with the power to make Federal statutes for specific purposes that is enforceable by action of the Federal Government. The legislatures of the States have the power to make laws on anything, unless prevented by the Federal Constitution or Federal statutes. For the United Kingdom, environmental legislation has a number of different sources as the United Kingdom is divided into three territories — England and Wales, Scotland, and Northern Ireland — each with its own legislation. In addition, Thornton and Beckwith (2004) state that around four-fifths of all environmental legislation in the United Kingdom now has its origin in the European Union (the "EU").

The following is a summary of the various sources of EU law, in terms of their status for legal decision-making in the United Kingdom (Wolf and White, 1997; Stookes, 2005):

i. *Treaties:* Treaties are the constitutional basis of the EU and represent the highest level of law;
ii. *Regulations:* Regulations are binding in their entirety and are applicable in all Member States;
iii. *Directives:* Directives are the primary means for implementing EU environmental legislation. Directives are binding on all Member States to the extent that they apply. Member States have the discretion as to the form and method of how the Directive is to be implemented;
iv. *Decisions:* Decisions have the force of law and are binding in their entirety on Member States, corporations and individuals; and
v. *Recommendations and Opinions. Plans and Policies, e.g. Environmental Action Programmes:* All of the above sources have no binding force but should be of persuasive value to national courts.

The ascendancy of EU law in relation to national law can be understood by considering the following principles, as reviewed by Wolf and White (1997) and Stookes (2005), which have emerged from EU case law:

i. EU law has precedence over national law and prevails where there is inconsistency between domestic legislation and EU law;
ii. Relevant provisions of a Directive can be enforced at the suit of individuals in national courts, provided that all the following conditions are satisfied: the content of the relevant provision is clear and precise, the relevant provision is unconditional and there is no room in the provision for the exercise of discretion by the Member State; and
iii. Where national legislation is open to interpretation, national courts should interpret that law, as far as practicable, to be in accordance with the wording and purpose of EU legislation.

INTERNATIONAL ENVIRONMENTAL TREATIES AND CONVENTIONS

Although environmental legislation has emerged as a significant source of legal rights, international treaties and conventions relating to the environment have also existed over time. International treaties and conventions represent another dimension which must be considered as a potential source of power for facilitating environmental negotiations. Multilateral international treaties, involving many nations as signatories, are referred to as conventions, e.g. the *Convention on Biological Diversity* signed at the United Nations Conference on Environment and Development at Rio de Janeiro in June 1992. Other international treaties may be bilateral as where only two nations are signatories to the treaty. International treaties have addressed areas of global environmental concern such as greenhouse gases, biodiversity, trans-boundary pollution, environmental impact assessment, world cultural and natural heritage, and hazardous chemicals.

International treaties and conventions move through a sequence of steps. The initial step is for a nation to enter into and to complete negotiations on the subject matter and to then become a signatory to the treaty or convention. Signature, by itself, does not result in the treaty becoming legally binding on the signatory nation. Nations that signed a treaty, when it was open for signature, can proceed to the next step — consenting to be bound by a treaty, through the alternative processes of ratification, acceptance or approval. These different processes arise because of the diversity of global legal systems in different nations, but nevertheless all result in a similar outcome. One further alternative process exists: accession. Where a nation did not sign a treaty during the time it was open for signature it can only accede to be legally bound by the treaty (Secretariat of the Convention on Biological Diversity, 2007a). The processes of ratification, acceptance, approval and accession are specific for the political process for each nation that consents to be bound. For example, for many nations the legislature (or parliament) or the head of state will decide whether to ratify a treaty or convention.

Ferrey (2004) identifies the following dilemma where a nation consents to be bound by the terms of the treaty or convention: if there is no way for a treaty or convention to be enforced, why should a nation comply? Ferrey provides the following reasons that may be relevant for compliance:

i. Consensual treaty agreements are entered into in the belief that it is in a nation's best interest;
ii. Where there is a global common interest to comply with the treaty, the expectation for a stable outcome will be facilitated; and
iii. Compliance with obligations under a treaty maintains a nation's trustworthiness, international honour, influence, prestige and good reputation.

Judicial Enforcement of International Treaties and Conventions

Kiss and Shelton (2004) state that international conventions and treaties often invade traditional spheres of functions of government through obligations contained in the terms of the treaty agreement that set out the principles or goals required for environmental management and protection. Treaties or conventions may be capable of immediate judicial application, i.e. "self-executing" provisions. This is not always the case and the treaty is referred to as being "non-self-executing". For example, the principles or goals of the treaty may be set out in general terms; alternatively, a signatory nation may wish to impose stricter national controls compared to the treaty itself. In addition, as a general rule, a treaty regulates relations between signatory nations rather than individual private parties. Because of these reasons, a ratifying signatory nation may be required to enact necessary domestic legislation to give effect to the terms of the treaty. In addition to achieve regulatory control as a violation by individuals of the treaty creates liability under the legislation and not the treaty (Craig, 2005).

In Australia, following ratification, the Federal parliament may enact legislation to implement a particular treaty or convention for it to become part of the domestic law and so have binding effect. A convention or treaty which has not been incorporated into domestic law cannot operate as a direct source of individual rights and obligations in Australia, unless its provisions have been validly incorporated into law by statute: *Minister of State for Immigration and Ethnic Affairs v Teoh* (1995) 183 CLR 273. Similarly, international conventions or treaties only become part of the national law, in the United Kingdom, following legislation enacted by parliament: *Blackburn v Attorney-General* [1971] 1 WLR 1037.

In the United States, many treaties foresee local enforcement by domestic law and have led to Federal statutes, as well as State legislation, being introduced to implement a particular treaty or convention, e.g. the *Convention on International Trade in Endangered Species in Fauna and Flora* and the *Endangered Species Act* and the *Mammal Protection Act* (Plater *et al.*, 2004). Craig (2005) considers that whilst the Constitution of the United States prevails over both Federal statutes and international treaties, the observation is made that Federal courts generally treat treaties like Federal statutes.

SCIENTIFIC/TECHNICAL FACTS AND THEORY

In contrast to questions of law, the facts are the steps along the way to determining the ultimate issue in dispute, as prescribed as a question of law. The facts, in themselves, are not the decision end-point in the legal process.

(a) Acceptance of Scientific Findings

The data and information upon which environmental conflicts are based are heavily founded on scientific, statistical and mathematical materials and have become increasingly sophisticated. Science relies on the give and take of criticism, testing, experimentation, peer review and publication to evaluate the validity of a scientific finding or theory. Contrary to a long-held misconception, science does not generate exact knowledge with logical certainty. Divergent scientific opinion on any issue will always exist. The central test employed by scientists to determine the validity of a finding, reasoning or theory, in any given context, is acceptance through widespread consensus (Black, 1988).

The consensus test of science is consistent with the approach taken by some courts in determining whether facts or theory contained in expert opinion evidence forms part of a body of knowledge which is sufficiently organised or recognised to be accepted as a reliable body of knowledge. A very good, early example of a court giving effect to the scientific method was the decision in *Frye v United States* 293 F. 1013 (1923) of the Court of Appeals of the District of Colombia in the United States, and its test based on the general acceptance standard. The adoption of a "general acceptance test" requires evidence, to be admissible, to conform to methods and principles which have received widespread acceptance in a particular field of knowledge. Frye's "general acceptance test" was superseded by the adoption of the Federal Rules of Evidence: Rule 702 [509 U.S. 579, 2].

In *Daubert v Merrell Dow Pharmaceuticals Inc.* 509 U.S. 579 (1993), the United States Supreme Court established an even stricter test for the judicial assessment of the reliability of scientific claims for the admissibility of expert opinion evidence. The Supreme Court determined that the trial judge was to act as a gatekeeper for scientific expert testimony and required to make a preliminary assessment of scientific claims, when Rule 702 of the Federal Rules of Evidence applied, to ensure that scientific evidence was "both relevant and reliable". The Supreme Court concluded that in relation to "*whether the testimony's underlying reasoning or methodology is scientifically valid and properly can be applied to the facts at issue* [that] *many considerations will bear on the inquiry*". These considerations include:

"[1] whether the theory or technique in question can be (and has been) tested;

> [2] whether it has been subjected to peer review and publication;
>
> [3] its known or potential error rate; and
>
> [4] the existence and maintenance of standards controlling its operation, and **whether it has attracted widespread acceptance within a relevant scientific community.** The inquiry is a flexible one, and its **focus must be solely on principles and methodology, not on the conclusions that they generate**" (Author's emphasis).

There would be little dispute that the approach of the United States Supreme Court in *Daubert's* case is consistent with the standards and criteria used by science for evaluating the reliability of a scientific finding or theory. That is the enduring criteria of testability, objectivity and impartiality, together with the test for acceptance of widespread consensus within the scientific community following peer review and publication (Black, 1988; Christie, 1991). Consequently, the application of the "Daubert standard" would be a relevant consideration in reviewing divergent scientific opinion where negotiation and ADR were used to resolve an environmental conflict. The "Daubert standard", effectively applied, represents a significant foundation for parties to a dispute to strengthen their bargaining power in environmental negotiations as it would facilitate knowledge power where information is a primary source of conflict.

Whilst the US legal principles have been considered in Australia and the UK, they have not been applied to the same extent. In Australia, in *Casley-Smith and Ors v FS Evans & Sons Pty Ltd and Anor (No. 1)* (1988) 49 SASR 314, Olsson J referred to authoritative texts and decided cases that established a number of propositions, including whether expert opinion evidence "*forms part of a body of knowledge ... which is ... sufficiently organised or recognised to be accepted as a reliable body of knowledge*" (at 320). In *Telstra Corp. Ltd v Hornsby Council* (2006) 146 LGERA 10, the Land and Environment Court made the following conclusion:

> "In Daubert v Merrell Dow Pharmaceuticals … the United States Supreme Court held that in a case involving scientific evidence, the evidence must pertain to scientific knowledge. The adjective 'scientific' implies a grounding in the methods and procedures of science and the word 'knowledge' connotes more than subjective belief or unsupported speculation. The requirement that expert evidence pertain to scientific knowledge establishes a standard of evidentiary reliability".

In the UK, the Australian focus on reliability of scientific evidence was also followed. In its consideration of *Frye* and *Daubert*, the Court of Appeal, in *R v Dallagher* [2002] EWCA Crim 1903, reviewed the English judicial approach in a number of cases from which it seemed to the court that an analogy with the United States Federal Rules of Evidence Rule 702 was clear, that is, the admissibility of a technique being dependent on it being generally accepted as reliable in the relevant scientific community. The court made the following observation:

> "As is said in the current ninth edition of Cross and Tapper on Evidence at 523 after a reference to Frye –
>
> The better, and now more widely accepted, view is that so long as the field is sufficiently well-established to pass the ordinary tests of relevance and reliability, then no enhanced test of admissibility should be applied, but the weight of the evidence should be established by the same adversarial forensic techniques applicable elsewhere".

(b) Proof of Facts: Legal and Scientific Concepts

There is a distinction between the civil standard of legal proof and the scientific standard of proof that needs to be recognised when evaluating data or information conflicts. Environmental conflicts that are litigated will generally involve civil actions rather than criminal prosecutions. The standard of proof required to discharge the onus (or burden) of proof, in a civil action, is characterised as being "*on the balance of probabilities*". The civil standard of legal proof has been commented upon by Lord Denning in *Miller v Minister of Pensions* [1947] 2 All ER 372 as:

> "It must carry a reasonable degree of probability but not so high as is required in a criminal case. **If the evidence is such that the tribunal can say: we think it more probable than not, the burden is discharged**, but if the probabilities are equal it is not" (Author's emphasis).

In civil actions, the question of satisfying the requirement of "more probable than not" may vary over a continuum depending on the degree of probability reaching 51 per cent through to a question of how short of 99 per cent would be required to establish proof. In contrast, the standard of proof in criminal prosecutions, namely "*beyond a reasonable doubt*", is far more onerous than the civil standard of legal proof and is reasonably similar to the scientific standard of proof. Science generally accepts that the standard scientific criterion for causality, in an animal or human study, would be founded on a 95 per cent (sometimes 99 per cent) confidence level. The following statement by an eminent English statistician — later knighted for his services to statistics — is the source of the protocol that is an enduring feature of scientific research:

> "If one in twenty does not seem high enough odds, we may, if we prefer it, draw the line at one in fifty (the 2% point), or one in a hundred (the 1% point). Personally, the writer prefers to set a low standard of significance at the 5% point,

and ignore entirely all results which fail to reach this level. A scientific fact should be regarded as experimentally established only if a properly designed experiment, rarely fails to give this level of significance (Fisher, 1926)".

A 95 per cent confidence level means that there would be a statistical likelihood in only five out of one hundred cases (or a one in twenty possibility) that the experimental finding for a causal link or association was due to mere chance. A confidence level in a study of less than 95 per cent (or 99 per cent) does not indicate there is no causal link or association but rather that there is an increased possibility that the finding is a result of experimental error or chance.

Based on the very clear distinction between the civil standard of legal proof and the scientific standard of proof, preparation for negotiation must be based on a rigorous examination of the scientific evidence for its reliability and validity to ensure that it is consistent with the standards and criteria of science for proof. Any failure to do so will limit any knowledge power arising from reliance on scientific data and information and will lead to a power imbalance for negotiation.

POLICY

A policy represents the general principles a Government agency uses to provide guidance to its public administrative decision-making functions. It is well recognised that a policy is not law and does not constitute a binding rule on anyone (Kubasek and Silverman, 2005). An exception does arise where such a requirement is prescribed in a statute: *Skoljarev v Australian Fisheries Management Authority* (1995) 22 AAR 335.

(a) United Kingdom

Consistency in public administration has been a principle of English law for at least one hundred years, having its foundation in fairness between citizens and a general presumption that Ministers will follow their policies unless there is reason not to do so (Lord Justice Sedley, 2004). As Lord Justice Sedley contends, policy today represents a key foundation of the public's expectations as to how Government will treat them. Important decisions made by the Executive branch of Government that are discretionary and outside the statutory framework should be accompanied by a policy to facilitate the integrity and consistency of the administrative decision-making process; any act of unfairness should be open to correction by judicial review. In addition, as Lord Justice Sedley points out, the reason why public law requires decision-makers to properly take everything relevant into account, including a policy, is that a decision that falls short of this standard is "*not only formally bad but likely to be unjust in its outcome*".

The public law requirements for procedural fairness (formerly referred to as "natural justice") in the UK (and Australia) for just and fair processes in legal proceedings, have now extended to incorporate a principle of "legitimate (or reasonable) expectation". In terms of the link between policy and procedural fairness, Lord Justice Sedley (2004) states that procedural fairness illustrates the sensitivity between policy needs and the interests of an individual. The power to change a policy is just as important as making it. Therefore, if a new policy, no matter how well made, fails to foresee the effects it may have on members of the public who have relied on the old policy, an arguable case for judicial review could be raised. A policy change that had unfairly frustrated the legitimate expectation that had been held under the earlier policy may be the basis of such a case.

(b) Australia

The Federal Court of Australia has also recognised that, in clearly defined circumstances, a policy can facilitate the integrity of the administrative decision-making process. Achieving this goal requires the presence of legal requirements that are embodied in the observations made by Brennan J (as he then was) in *Re Drake and Minister for Immigration and Ethnic Affairs (No. 2)* (1979) 2 ALD 634. Specifically, the policy must be consistent with the statute and "*must allow the Minister to take into account the relevant circumstances, it must not require him to take into account irrelevant circumstances, and it must not serve a purpose foreign to the purposes for which the discretionary power was created*".

(c) United States

A similar situation to Australia and the United Kingdom also applies in the United States in that a "general statement of policy" has no binding effect on the public or the court. Under the *National Environmental Policy Act of 1969,* the coordination of environmental policy is solely under the control of the Environmental Protection Agency (Kubasek and Silverman, 2005). A general statement of policy only states what an agency seeks to establish as a policy. It is exempt from notice and comment procedures for rule-making under the *Administrative Procedure Act* [APA 5 U.S.C.A., Section 553(b)] and is not judicially enforceable against an agency (Pierce, 2002). The observations of the court in *Pacific Gas and Electric Co. v Federal Power Commission* 506 F.2d 33 at 38 (D.C. Circuit 1974) explain the status of a general statement of policy:

> "A general statement of policy is the outcome of neither a rulemaking nor adjudication; it is neither a rule nor a precedent but is merely an announcement to the public which the agency hopes to implement in future rulemakings or adjudications".

As Pierce (2002) points out, the distinction between a general statement of policy and a legislative rule may be difficult. Courts in the United States distinguish between legislative rules and a general statement of policy, depending on whether the policy statement binds members of the public. A general statement of policy that has a binding effect could be considered as a rule rather than a general statement of policy (Pierce, 2002). However, the quandary for decision-making for an agency, in these circumstances, has been clearly identified by Pierce: if a general policy statement by an agency is found by the court to be binding in some important respect then it can only be proclaimed through the use of rulemaking procedures under the APA. As for any legislative rule, it could be subject to judicial review. On the other hand, if the court concluded that the statement was a general policy statement, an agency would not be permitted to bind members of the public.

VALUES

Environmental values are a common source of conflict when environmental problems arise. All forms of scientific knowledge are grounded in some value position — ideological or philosophical. The concept of values is multi-disciplinary, having its foundations in philosophy, sociology and economics. The different ideologies and philosophies that exist within the community reflect the relationship between humans and their environment.

The central axis for values is said to be one of "ecocentrism" (or earth-centred) and "anthropocentrism" (or human-centred). Within the axis, a subdivision for the characteristics of each environmental ideology can be further defined to provide some understanding of the values that humans may hold with respect to the environment. In turn, these characteristics may reflect their attitudes towards consensual decision-making and compromise — key elements for negotiating an environmental conflict. These characteristics have been summarised as follows (O'Riordan, 1989a; Gough *et al.*, 2000):

- *Gaianism* — trust "Gaia" (or faith in the rights of nature and the essential need for the co-evolution of human and natural ethics) above all;
- *Accommodation* — value "life" above all;
- *Communalism* — value "fairness" above all; and

- *Intervention* — value "progress" above all.

(a) Values and the Courts

For environmental values to have any weight in litigation, the value must be prescribed in a statute as a matter that is required to be taken into account by the court as a legal obligation. Such a legal obligation can arise expressly when the words of the statute identify a specific value as such a matter to be considered. Alternatively, a legal obligation can arise impliedly, when such a matter may legitimately be inferred from the words of the statute. It is not enough that the value is a consideration which many people, including the court itself, would take into account if they had to make the decision: see *CREEDNZ Inc. v Governor-General* [1981] 1 NZLR 172.

An example of the court's approach to consider values — in this case, the moral ground that deer-hunting was "*unacceptable and unnecessarily cruel*" — arose in the English case *R. v Somerset County Council Ex Parte Fewing & Others* [1995] 1 WLR 1037. A challenge was made that the administrative decision by the County Council to ban hunting of deer on public land was unlawful because it was based on moral grounds, rather than in accordance with the statutory powers under the relevant statute (the *Local Government Act 1972*). In his reasoning, the presiding member of the Court of Appeal, Sir Thomas Bingham, Master of the Rolls, stated (at 1042):

> "The court has no role whatever as an arbiter between those who condemn hunting as barbaric and cruel and those who support it as a traditional country sport more humane in its treatment of deer ... than other methods of destruction such as shooting, poisoning ... This is, of course, a question on which most people hold views one way or the other ... the question ... which the court is obliged to answer ... is whether the county council acted lawfully in making the decision it did on the grounds it did ... [whether the council was entitled] to give effect to their acceptance of the cruelty argument."

The Appeal Court referred to the legal principle in *CREEDNZ Inc.* as "salutary" and decided (in a 2:1 majority) that the County Council had not been lawfully entitled to make a decision to impose a ban on deer-hunting on moral grounds as it did not fall within the powers that the Council could exercise under the statute. Moral grounds for the hunting of deer on public land were not a consideration required to be taken into account by the court as a matter of legal obligation under the *Local Government Act.*

Where values are a source of conflict, the legal principle from New Zealand case law in *CREEDNZ Inc.* is useful in understanding their status for decision-making in litigation. Whilst the principle is only of persuasive value for courts outside New Zealand, it has been applied in a number of cases by courts in England and Australia.

(b) Incorporating Values into the Environmental Conflict Resolution Process

Two value-related interventions identified by Moore (1986) to offset conflicts over values are "*to avoid defining the problem in terms of values*" and "*to search for superordinate goals that all parties share*". An Australian public inquiry established by the Federal Government (Commission of Inquiry into Shoalwater Bay, Capricornia Coast, Queensland, 1994a) considered these interventions by placing a special focus on shared environmental interests as the higher ranking goal. The study area was the Shoalwater Bay Military Training Area. The existing and proposed land uses that were the subject of the public inquiry can be summarised as follows:

- The study area included both land and sea (including parts of the Great Barrier Reef Marine Park) and occupied 520,000 hectares. The study area was owned by the Commonwealth of Australia and was extensively used by the Department of Defence for military training purposes and therefore significant in terms of national security. Field exercises involved the Australian Defence Forces alone or in combination with defence forces from the United States and South-East Asia, as part of the ANZUS International Treaty.
- A large part of the study area was of national significance for its natural and cultural heritage values and was recognised as such by being placed on the Commonwealth's "Register of the National Estate". The coastal part of the study area was the sole water supply catchment for residents living in the local area south of the study area. Aboriginal interests had a past connection with the study area, although they no longer lived in the study area.
- Commercial fishing occurred in the waters of the study area. Mining leases for mineral sands (rutile, zirconium, titanium) were held but no mining had commenced. In addition, an application had been made for further exploration for mineral sands in the study area.

Methodology

Divergent opinion exists as to the relevance of value and attitude surveys in public interest environmental conflicts. For example, public opinion surveys are of limited use because of the many sources of experimental error that restrict the confidence levels and reliability of the findings, e.g. sampling methods, questionnaire design, data analysis and background briefing of those surveyed. The alternative viewpoint would be that a properly constructed and analysed survey into values and attitudes would be the ideal procedure to determine public opinion on questions of public support and beliefs in public interest environmental conflicts — especially where a random stratified survey was undertaken on a local, regional and state-wide analysis.

There were four stages in the process for incorporating values into the public inquiry's decision-making process:

Stage 1: a value and attitude survey A survey into community values and attitudes in relation to acceptable land uses for the study area was undertaken during the course of the public inquiry. There were ten existing and proposed land uses that had been defined in the Terms of Reference of the public inquiry.

Stage 2: identification of community interests and environmental values The strategic purpose of the survey was to identify the interests each party had that reflected their values and attitudes towards the environment and the existing and proposed land uses, rather than simply to base a decision on the community support measured in the survey. A response was sought on the importance of eight defined environmental values such as ecological values or economic values. The values where strong community support was demonstrated were:

i. "Protection of the water supply for the Capricorn Coast";
ii. "Protection of plants, animals and environmental values"; and
iii. "Defence use of the study area".

Stage 3: evaluation of community interests and environmental values The Commission of Inquiry then engaged external, independent scientific experts to provide objective scientific opinion on future impacts on environmental values that had been identified in the survey as significant sources of concern. This step enabled an assessment to be made on whether the basis for the community responses was either "hypothetical" or "real" for each value identified. Public submissions were then sought in relation to the expert scientific opinion obtained before any findings were made.

Stage 4: the outcome The survey showed a remarkably consistent State-wide pattern of responses with respect to both (i) acceptable land use and (ii) place of residence in the State of Queensland for those surveyed (Figure 2.2). The survey results clearly identified that the most favoured option for the future use of the study area was for its conservation (88 per cent acceptable). A further grouping of land uses identified as acceptable were scientific research, defence activity, nature-based tourism, recreation and Aboriginal use.

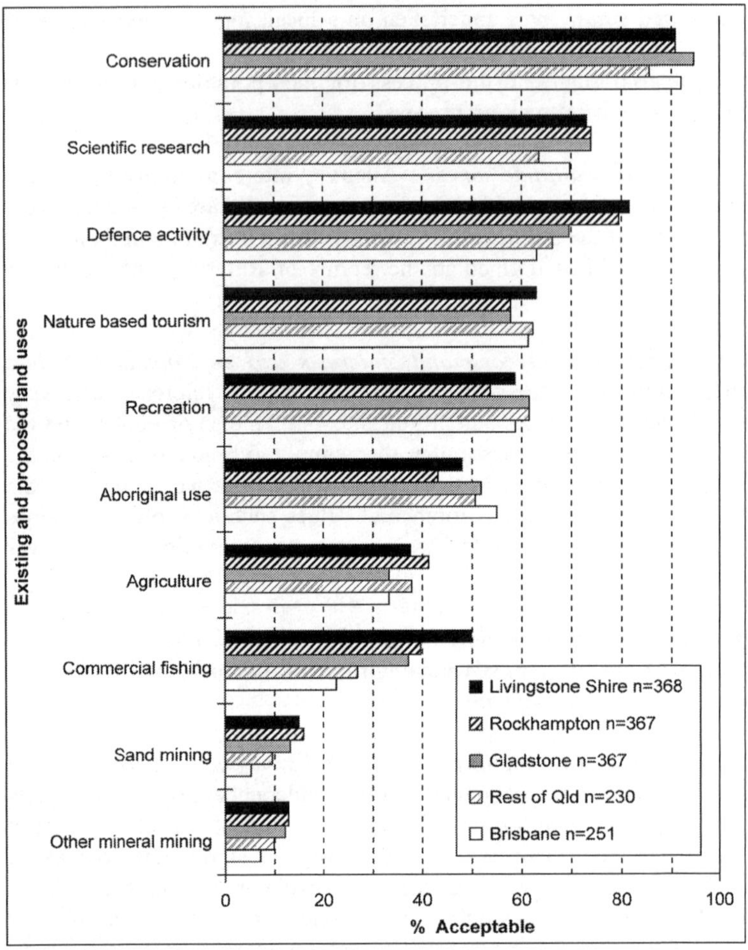

Notes:
i. Livingstone Shire (or local authority) represents the local area where the environmental impacts of proposed land uses would be incurred.
ii. Rockhampton represents a Shire in the regional area immediately adjacent to Livingstone Shire.
iii. Gladstone represents another Shire in the regional area, 100 km south of Livingstone Shire where the mineral sands would be processed; economic benefits would accrue but environmental impacts from the proposed land uses would not be incurred.
iv. Brisbane is the capital city of the State of Queensland ("Qld"), 800 m south of Livingstone Shire.

Figure 2.2 Incorporating values into the environmental conflict resolution process. Stratified sampling of residents on a local, regional and State-wide basis, Queensland, Australia. (AGB McNair and Commission of Inquiry into Shoalwater Bay, Capricornia Coast, Queensland, 1994)

The final recommendations made by the Commission of Inquiry to the Federal Government reflected the need for an existing or proposed land use activity to not degrade the environmental values. For example, in terms of the most favoured option for a future land use, conservation, one key finding made by the public inquiry that was consistent with community opposition to land uses that might degrade the natural and cultural values was: "*The study area's terrestrial and marine environment represents large and relatively undisturbed habitat areas for significant floral and faunal assemblages, including populations of rare and threatened species ... the study area has a high level of biodiversity ...* ". Recommendations were made to support the findings by proposing measures for how the acceptable land uses of conservation and defence activity could co-exist and not degrade the values of "*protection of plants, animals and environmental values*" and "*defence use*" (Commission of Inquiry into Shoalwater Bay, Capricornia Coast, Queensland, 1994a).

FACTS AND KNOWLEDGE POWER

The scientific data and information base in an environmental conflict will invariably consist of (i) experimental facts, (ii) professional opinion through to (iii) value judgements. Given that scientific and technical facts will be both complex and numerous, and that scientific uncertainty will prevail in any environmental conflict, knowledge of facts provides a significant source of power for influencing the form a negotiated solution may take. The dilemma for conflict resolution is simple: understanding scientific data and information is a source of knowledge power; lack of knowledge power is an obstacle for effectively participating in environmental negotiations.

(a) Evaluating Scientific Information

The relevance and reliability of scientific data and information is dependent on scientific findings and conclusions — or theory — and reflects the need for their consistency with the scientific criteria of testability, objectivity and impartiality (Black, 1988). This step requires an evaluation of the underlying concepts, experimental methods and design, statistical methods and data analyses before any weight can be attached to the conclusions (Christie, 1991). The potential difficulty for lawyers, untrained in science, is to be able to pursue this sequence in order to evaluate information conflicts for environmental problems. In addition, because of the very clear distinction between the civil standard of legal proof and the scientific standard of proof, preparation for negotiation must be based on a rigorous examination of the scientific evidence for its relevance and reliability. Any failure to do so may create limitations during negotiations, by making a party vulnerable to the knowledge power of the other side.

Conclusions on the scientific database should be based on objective facts and not be open to a challenge on the grounds of subjectivity, e.g. a value judgement. Additionally, conclusions should not be hypothetical or speculative, i.e. the evidence is not real. A prerequisite for joint fact-finding is for the scientific data and information to be evaluated using objective criteria so that regardless of who evaluates the scientific database, the outcome will be the same. This goal can be achieved by:

- Selecting end-points for the environmental impacts that can be measured;
- Ensuring that a scientific data and information base is available to assess the end-points; and
- Relying on objective criteria to evaluate the scientific information base.

(b) Evaluating Potential Environmental Impacts

Environmental conflicts frequently involve *potential* impacts on the environment or on human health. In these circumstances, questions of causation cannot be evaluated directly as the event has not occurred. The approach used by science and law to assess facts and questions of causation, indirectly, is by the method of inference. The reasoning of LJ Wright in *Caswell v Powell Duffryn Associated Collieries* [1939] 3 All ER 722 at 733 is particularly relevant with respect to understanding the common law evidentiary requirements that apply for an inference to be established:

> "There can be no inference unless there are objective facts from which to infer the other facts which it is sought to establish … But if there are no positive proved facts, from which any inference can be made, the method of inference fails and what is left is mere speculation or conjecture."

For example, sampling is a scientific method that will be central to almost all environmental conflicts. Standard scientific methods for investigating environmental problems in the field, such as the concentration of a pollutant, animal numbers, species diversity and social surveys, all rely on sampling of only part of an area or population. To rely on the method of inference and to be able to make conclusions for the entire area or population, based on a sample of only part, the sample taken must be shown to be representative of the entire area or population that represents the spatial boundaries of the conflict. In order to establish that the sample is representative of the entire area or population, positive facts would need to be proved. Key facts in this regard would be to establish that the sampling plan was based on accepted scientific principles contained in standard sampling protocols, e.g. randomness, freedom from bias and the adequacy of the number of samples. Apart from sampling, another example in environmental management and protection where the method of inference is a central requirement relates to proof of causation and exposure to hazardous chemicals and public health problems.

AVAILABILITY OF FACTUAL INFORMATION: LITIGATION V NEGOTIATION

There are major differences between litigation and ADR related to the availability and evaluation of information. There is a very clear distinction in how information becomes available and exchanged in (i) litigation and (ii) ADR when it is part of the court-annexed ADR model and (iii) non-court-annexed ADR undertaken external to and independently of the courts.

(a) Litigation and Court-Annexed ADR

In support of lawyers and litigation, it should be emphasised that, notwithstanding the fact that most lawyers are not scientists, they do have a significant role, relative to scientists, in conflict resolution. In preparing for litigation, lawyers have available to them the processes of "discovery" and "interrogatories" (written questions submitted to the opposing party as part of discovery) that are under the control of the court. During discovery, lawyers identify and collect all documents, data and information that relates to the conflict, prior to the hearing. In this regard, the role of lawyers is to effectively organise extensive quantities of data and to arrange it logically in a manner consistent with evidentiary requirements, so as to ensure a party's case is put forward to its best advantage.

The processes of discovery and interrogatories and the organisation of extensive quantities of data and its logical arrangement by lawyers in litigation also provide the foundation for information disclosure when ADR is undertaken under the court-annexed ADR model.

(b) Non-Court-Annexed ADR

Non-court-annexed ADR refers to ADR undertaken independently of the courts and so unconnected and unrelated to any litigation commenced in the courts. Compared with courts, where relevant information is exchanged and assessed by the parties prior to being examined at the hearing, non-court-annexed ADR relies on a voluntary exchange of information. Some concerns have been expressed when ADR is undertaken external to and independent of the courts, in particular the absence of judicial control to prevent any abuse of the disclosure of information as there is no discovery process in these circumstances. Nolan-Haley (1992) raises an ethical issue in this situation in relation to truthfulness. Specifically, how much information is required to be revealed prior to commencing and during negotiations? It has been claimed that when negotiations are undertaken as part of the ADR process, it is possible for ethical norms to be violated in the belief that untruthfulness — or even misrepresentation — may not be discovered or punished (White, 1980). In the case of lawyers and legally accredited mediators, it needs to be emphasised that their codes of professional conduct require that they not be a party to any misrepresentations or misleading and deceptive conduct.

These concerns highlight the need to have some form of process equivalent to that of discovery in litigation, when ADR and negotiation is undertaken external to and independent of the courts. For example, by incorporating data (or information) mediation as a process of information exchange as part of the preparation phase for ADR before negotiations commence (Susskind and Weinstein, 1980).

(c) Data (or Information) Mediation

Data (or information) mediation is a structured process of information exchange between the parties and is the preliminary step in conflict management. The purpose of it should be seen in terms of its goal: the ADR equivalent of the legal process of discovery by ensuring there has been a full disclosure of scientific and technical information prior to negotiations commencing. Data mediation ensures that all information relating to the environmental conflict is fully disclosed in a manner procedurally fair to the parties. Information that is equally available to all parties ensures that negotiations can proceed on the basis that no cards are held under the table by any party (Christie, 1993a).

EXPERT EVIDENCE AND LITIGATION

Divergent scientific opinion on disputed issues in environmental conflicts will always exist within the scientific community so that courts will be heavily dependent on the opinion evidence of scientific experts. The legal principles for the role of an expert witness are well recognised and include that the expert cannot give opinion evidence as to the ultimate issue that is to be decided by the court. In addition, expert evidence presented to the court should be the independent product of the expert, uninfluenced as to form or content by the exigencies of litigation: LJ Wilberforce in *Whitehouse v Jordan* [1981] 1 All ER 267 at 276. In terms of evaluating the relevance and reliability of factual evidence, cross-examination has been referred to as "the greatest engine for truth" (Wigmore, 1983). One of the main arguments in favour of adversarial litigation is that it represents the most effective way to test the credibility of the evidence of a witness through the use of a skilled advocate, for each side, investigating the facts from a partisan perspective. However, the adversarial method of litigation has been criticised by the scientific community for the following reasons (Selinger, 1984):

i. Cross-examination is anti-scientific, e.g. by forcing answers to complex questions in simple terms and where it requires conclusions to be reached before all facts are assembled;
ii. Scientific opinion becomes polarised;
iii. The ease with which well presented, but erroneous science is accepted by the court;
iv. The fact that scientists cannot put their questions and arguments to each other directly in court, but only through partially comprehending lawyers; and
v. A successful attack on a scientist's work on a single point can be devastating. Even if the point in question is not destructive of the conclusion, it may be destructive of the credibility of the scientist.

A long-standing problem for expert evidence relates to the potential for bias by scientific expert witnesses, as is illustrated by the following judicial comment made by an English Court over 160 years ago: "*Hardly any weight is to be given to the evidence of what are called scientific witnesses: they come with a bias on their minds to support the cause in which they are embarked*": Per LJ Campbell, *The Tracy Peerage* [1843] 10 Cl & Fin 154 at 191.

A similar, contemporary observation on the nature of expert evidence was made by Weinstein J. (1986), a United States judge who presided over the "Agent Orange" lawsuits when they were sent to the Eastern District of New York in 1979: "[a]*n expert can be found to testify to the truth of almost any factual theory, no matter how frivolous*".

However, lawyers have contributed to this problem through their procedures for briefing expert witnesses. The following statement made in 1877 by the eminent jurist, Sir George Jessel, Master of the Rolls, in *Thorn v Worthing Skating Rink Co.* [1877] 6 Ch D 415, clearly shows how lawyers involved in the legal process may shop around for a favourable expert — which in itself may lead, involuntarily, to the expert becoming biased:

> "A man may go, and does sometimes, to half-a-dozen experts. I have known it in cases of valuation within my own experience at the bar. He takes their honest opinions, he finds three in his favour and three against him; he says to the three in his favour, will you be kind enough to give evidence? And he pays the three against him their fees and leaves them alone; the other side does the same ..."

Problems associated with the potential bias of expert witnesses in litigation have been addressed, over time, by courts. The use of court-appointed experts is one such approach where an issue is the subject of expert opinion and where the parties consent. Another process introduced in Australia by the Federal Court and the Administrative Appeals Tribunal is "*concurrent evidence*" (sometimes referred to as the "hot tub").

(a) Litigation and the Evaluation of Expert Evidence: Concurrent Evidence

Concurrent evidence, as a process for evaluating divergent expert opinion evidence, was introduced by the Federal Court of Australia to enable "cutting edge controversies" in big cases to be resolved. The rationale for concurrent evidence is to evaluate divergent expert evidence using procedures that would facilitate experts to debate the key issues in a way in that the judge and counsel could understand. The concurrent evidence process has been extended in the Commonwealth of Australia's Administrative Appeals Tribunal to "ordinary cases" in which divergent expert opinion evidence must be evaluated as part of the Tribunal's fact-finding process. The essential features of the concurrent evidence process have been described by Downes (2004) and include:

- To enable the reliability and validity of the evidence and opinions of experts to be better tested by the court or tribunal, legal representatives and other experts, with the aim of the evidence being comprehensively explained, understood and analysed;
- To assist experts in fulfilling their role as independent advisers whose primary role is to assist the court or tribunal; and
- To enhance the operation of court or tribunal proceedings by reducing the time taken to resolve matters.

Following the examination of the evidence of the experts at a hearing, the reasoning and opinion expressed in their written reports and oral evidence are tested by way of a panel session called a "hot tub". During the "hot tub" phase, each expert is invited to make a presentation that addresses their statements and to identify the key issues. The experts may consult, be challenged or discuss their views with the other experts on the panel. The court, or tribunal, may also ask questions of the experts as necessary. Finally, counsel for the parties are given the opportunity to ask questions of the experts in relation to any matters raised during the "hot tub" interchange as well as on documents and materials filed with the court as part of the process of discovery.

(b) ADR and the Evaluation of Expert Evidence: The Scientific Round Table

The scientific round table is a structured process for evaluating divergent viewpoints on scientific and technical issues in environmental conflicts. It has been developed and used by the author for managing information conflicts, where negotiation is undertaken, external to and independent of the courts, to resolve environmental conflicts. The scientific round table adopts a joint fact-finding approach for evaluating the scientific issues in dispute, based on a full and fair disclosure of all relevant information. Joint fact-finding at the scientific round table overcomes the obstacle of polarised scientific opinion arising from the adversarial process and litigation. The representatives at the scientific round table are scientific professionals, having expertise in the subject matter of the conflict and who have been nominated by each party to the conflict as their representative at the round table. An independent dispute resolver convenes the scientific round table.

The scientific round table has two objectives for conflict management. The ADR process used for achieving both objectives is independent expert appraisal. Where ADR is undertaken independently of a court and there is no formal process of discovery and information exchange, data (or information) mediation, convened by the dispute resolver is the first objective. The purpose of data mediation is to ensure that all relevant scientific and technical information known to be published and which is relevant to the issues in dispute, is disclosed and made available through information exchange. The dispute resolver and the scientific experts all have a role in identifying information that needs to be the subject of information exchange and so become the common database made available to all parties.

The second objective of the scientific round-table process is for the scientific experts to reach consensus on each disputed scientific and technical issue. The round table gives effect to one of the key elements of principled negotiation by insisting that agreement on disputes issues be based on objective criteria (Fisher *et al.*, 1991). Consequently, a common set of objective criteria to evaluate the common database is determined by the scientific experts. Following review of the common database, each scientific expert at the round table provides an overview of the key scientific and technical issues in dispute. As the round table is structured as a joint fact-finding exercise, the scientific experts cooperatively interact with one another in conferring or challenging differing perspectives held on disputed issues. The dispute resolver may also question the experts. Where consensus cannot be reached on any issue, the dispute resolver draws on information provided by the scientific experts at the round table and, where appropriate, may undertake additional research, in order to provide a non-binding opinion to the experts on each disputed issue. The purpose of this non-binding opinion is to facilitate the experts to reach consensus on the disputed issue in question through further debate and discussion.

The conclusion of the scientific round table is reached when the experts reach consensual agreement on the disputed scientific and technical issues. Where consensus has not been reached on any disputed issue, the non-binding opinion of the dispute resolver on the issue is included as part of the outcome. The dispute resolver prepares a summary of outcomes from the joint fact-finding of the common scientific database at the scientific round table. The summary of outcomes becomes the foundation for negotiating agreement during the conflict resolution stage.

LITIGATED V NEGOTIATED OUTCOMES

All environmental conflicts will involve the same elements: questions of law, facts, policy and values. However, the status and role for each element in arriving at litigated or negotiated outcomes can be distinguished. These features need to be recognised in considering a conflict resolution pathway as apart of a BATNA analysis.

(a) Status of Elements

The outcome of litigation is for the disputed factual issues to be decided on the basis of a question of law. In contrast, the basis for arriving at a solution by ADR and negotiation can be one, or more, of the following elements, as agreed to by the parties, at the outset before negotiations commence: relevant and reliable scientific evidence, law, policy, values and ethics.

(b) The "Rights-Based" Approach to Conflict Resolution

Litigation and the legal process, based as they are on law and legal rights, have been described as a "*rights-based approach*". The legal process relies on the operation of the doctrine of precedent by applying legal principles arising from common law to ensure consistency in judicial decision-making in similar fact cases. In contrast, there is no precedent set by any agreed solution where a conflict has been resolved by ADR and negotiation.

ADR can also be rights-based in specific circumstances e.g. where the interpretation of the law is not in dispute. If ADR is undertaken as part of a court-annexed ADR system, the usual requirement is for any agreement to be consistent with the law before a consent order that is enforceable as a court order will be made. In addition, where non-court-annexed ADR is undertaken independently of and external to the courts, the parties may voluntarily agree, at the outset, for any negotiated agreement to take into account legal rights.

The problem for negotiating a solution using a rights-based approach is that a party's perceptions of decision-making may well be in accordance with what they believe will be the likely remedy should the dispute be litigated. A rights-based approach focuses on an adversarial, rather than a cooperative dynamic, with an emphasis on the immediate dispute rather than resolving the underlying conflict (Duncombe and Heap, 2004). Furthermore, an inordinate emphasis in a rights-based approach encourages "*positional bargaining*", where the object is either to win, to maximise benefit, to give less to the other parties or to avoid compromise; the positional negotiator tries to persuade the other parties to the negotiation that their solution or demand is the best of all possible solutions (Duncombe and Heap, 2004).

(c) The "Interest-Based" Approach to Conflict Resolution

There is an alternative approach to the rights-based approach and the difficulty of positional negotiation. The "*interest-based negotiation*" approach offsets the problems associated with positional negotiation by focusing on the interests of the parties to the dispute that must be satisfied in order for a solution to be found. The underlying rationale for interest-based negotiation is well established under the concepts of principled negotiation of Fisher *et al.* (1991). The dynamic is one of negotiation providing the parties with an opportunity for joint problem-solving rather than for one side to win. The objects of the interest-based negotiator include:

i. To find a mutually acceptable solution for the problem that results in mutual gain, e.g. by exploring all possible solutions until the best solution becomes apparent and can be agreed to; as well as
ii. Establishing the underlying needs and concerns of each party and to find ways to satisfy these interests cooperatively.

Interest-based negotiation involves a joint fact-finding and joint problem-solving approach for finding mutual agreement. Interest-based negotiation is a significant and distinguishing characteristic of ADR and negotiation, but it is not a feature of litigation.

(d) Values and Conflict Resolution

Where values or morals or ethics are a significant source of conflict, the key question is whether to litigate or negotiate. Gusman and Harter (1986) contend that conflicts requiring resolution of divergent opinions on moral values are inappropriate for negotiation. Moreover, Harter (1982) considers that where basic values would be compromised, agreement would be unlikely. However, where values, morals or ethics have not been expressly or impliedly provided for in a statute, they cannot be treated as statutory law and so are not matters to be taken into account by a court as a legal obligation. In this situation, litigation would be ill-advised. This leaves ADR and negotiation as the best alternative, but subject to the qualification that value-related interventions as identified by Moore (1986) can be effectively applied.

3. Constraints to participation in public interest environmental conflicts

"If **Donoghue v Stevenson** [the basis for the modern law of negligence] had arrived in 1992 instead of 1932, the case would never probably have got to court. In 1932, it was a privilege to argue a case for no fee. In 1992, it would be refused legal aid."

Selby (1992)

"Bernard claims that the citizens of a democracy have a right to know. We explained that, in fact, they have a right to be ignorant. Knowledge only means complicity and guilt. Ignorance has a certain dignity. I remarked that one does not just give people what they want, if it's not good for them. One does not, for instance, give whisky to an alcoholic."

Sir Humphrey Appleby in Lynn, Jonathan and Antony Jay (1981), *The Complete Yes Minister*, Chapter 1, 'Open Government', BBC Books, London

" ... the environmental decision-making process [in the Canadian Arctic now incorporates] local values, priorities and traditional knowledge in environmental research and management. Especially important has been the emergence of traditional environmental knowledge as a mechanism by which participatory approaches can be implemented."

Berkes *et al.* (2001)

" ... environmental issues are best handled with the participation of all concerned citizens at the relevant level."

Principle 10, Rio Declaration (1992)

INTRODUCTION

Public interest environmental conflicts generally arise because of a community concern to protect the environment and not for the purpose of obtaining some form of personal material benefit. The role public participation brings to public interest environmental conflict resolution includes maximising the information available to the decision-maker to ensure that any decision by a Government agency to approve a proposed project or activity is made with a full understanding of the issues and consequences of the proposal (Wilcox, 1985). The importance of this role needs to be considered in the context that information conflicts are a primary source of conflict for environmental problems.

Today, the opportunity for the community to be involved in public interest environmental conflicts is now very much a global norm. Processes for involvement exist in a number of forms. The dilemma for the community involved in any one participatory process is knowledge of their procedural rights and the level of power they have in relation to the outcome. Abaza *et al.* (2004) give a clear analysis and distinction between two of the most commonly used participatory processes:

- *Community consultation*: A two-way process of information exchange between the Government agency, and/or the proponent, and the community. Where community comment is sought the Government agency and/or the proponent is not bound by the views expressed in its decision-making process; and
- *Public participation*: Shared involvement and responsibility is a central feature as public participation implies an element of joint analysis and control over decisions and their implementation. Mediation and consensus-building may be an agreed part of the participatory decision-making process.

An additional process for involving the public is community engagement. Whilst this concept has existed for some time, there is no generally agreed definition. What is accepted is that the concept of community engagement incorporates a variety of approaches, such as consultation and participation, to involve the community and to provide the community with a greater role in collaborative decision-making. Notwithstanding the benefits that public participation brings to the process for resolving environmental conflicts, there are a number of significant constraints. The availability of financial resources for the costs of expert scientific reports, as well as legal costs, will always be a problem for community groups compared with Government agencies or developers. In this regard, financial resources frequently represent a major obstacle for the power the community has to effectively participate in resolving environmental conflicts. However, such a perception tends to deflect from other constraints that may restrict the effective participation of the community in finding solutions for public interest environmental conflicts.

The principles for "*effective public participation*" have been an accepted body of scientific knowledge for a considerable period of time, e.g. the

Institute for Participatory Planning (1981). One such principle defined by the Institute requires that "*information should be freely available to all participants*". This is an extremely significant concept that lies at the heart of resolving public interest environmental conflicts. A social survey to evaluate principles for "Public Participation Procedures for Environmental Decision-Making" was undertaken as part of an Australian public inquiry (Commission of Inquiry into Fraser Island and the Great Sandy Region, 1991) into a public interest environmental conflict — primarily over World Heritage Listing and logging of native forests on the largest sand island mass in the world (Fraser Island, 1630 km^2 in area, 123 km long and 25 km at its widest). The survey clearly established that, by far, the most important principle for public participation was that "*Communities and other interested parties should be given adequate, readily intelligible information on which to make decisions*".

This principle for effective public participation may be constrained by Freedom of Information ("FOI") legislation as information may not always be freely available for the public participation process. Accessibility to information held by Government agencies is dependent on the regulatory controls imposed by FOI legislation so that documents or information sought may not necessarily be disclosed by Government. Lack of access to information or documents held by Government will limit the power of the community to effectively participate in conflict resolution, regardless of the form of involvement provided. A further principle for effective public participation of the Institute for Participatory Planning (1981) requires the community to be "*aware of the level of power being offered*" to them in the participation process. For example, the level of power available may restrict the community to comment on a proposal by making written submissions, or the community may only have power to make recommendations as to the form of the final decision.

A more recent constraint to public participation that emerged in the 1980s in the United States is a tactic aimed at silencing private citizens and the community who oppose a development proposal or activity. The term "Strategic Litigation against Public Participation" ("SLAPP"), a tactic associated with the threat of legal action in environmental conflicts, had its origin in the United States in 1988 in an article published by two environmental lawyers — Professors George Pring and Penelope Canan. The tactic has now moved from the United States to other countries. SLAPP, as a tactic, is a source of coercive power to silence public opposition to a development proposal or activity. However, any such advantage in power may be offset if SLAPP is recognised as a tactic and is able to be countered.

INTERNATIONAL CONVENTION: PUBLIC PARTICIPATION

A landmark international convention for public participation was the *Convention on Access to Information, Public Participation in Decision-Making and Access to Justice in Environmental Matters* (the "Aarhus Convention") that became open for signature at Aarhus, Denmark in June 1998. The Convention was sponsored by the United Nations Economic Commission for Europe ("UNECE") and was open for signature by the 55 members of the UNECE, which includes all of western Europe as well as the United States, Canada and former states of the former Soviet Union (Kiss and Shelton, 2004). The United States is a non-signatory UNECE member to the Convention and is virtually alone amongst UNECE states in not having signed this Convention (Brunnée, 2004). Australia is not a member of the UNECE and so is ineligible to be a signatory to the Aarhus Convention.

Procedural Rights under the Aarhus Convention

The Aarhus Convention provides a comprehensive framework for procedural rights for public participation in environmental matters. Article 6 of the Aarhus Convention prescribes procedural rights for "*Public Participation in Decisions on Specific Activities*". Annex I of the Convention lists a broad range of 19 activities that apply to Article 6, e.g. chemical and nuclear installations, waste management and disposal and waste-water treatment plants, groundwater abstraction and construction of overhead power lines. An activity that is not listed comes within the scope of the Convention "*where public participation is provided under an environmental impact assessment procedure in accordance with national legislation*": Annex I, paragraph 20. Additionally, the provisions of Article 6 apply to decisions on proposed activities not listed in Annex I "*which may have a significant effect on the environment*": Article 6(1)(b).

Procedural rights for public participation under Article 6 of the Convention can be summarised as follows:

- The public are to be informed in a timely and effective manner of the environmental decision-making procedure in terms of: (i) a proposed activity, (ii) the nature of possible decisions or the draft decision and (iii) the public agency responsible for making the decision;
- Information is provided on the envisaged procedure in relation to: (i) the opportunities for public participation, (ii) the time and venue of any public hearings and (iii) an indication as to the available environmental information that is relevant to the proposed activity;

- A requirement for each party to give to the public access to information free of charge, and all information relevant for environmental decision-making as soon as it becomes available;
- Provision for early public participation as well as reasonable timeframes provided for the public to prepare and to participate effectively during the different stages of the participatory process; and
- Procedures for public participation allow comment, information, analysis or opinion considered by the public to be relevant to the proposed activity. Such inputs may take the form of written submissions or at a public hearing or inquiry.

PUBLIC PARTICIPATORY APPROACHES

(a) Australia

The original model relied on by Government, as a participatory approach to maximise public participation, was the Commission of Inquiry (or "public inquiry"). A Commission of Inquiry acts as a springboard for Government decision-making, especially where the public interest environmental conflict is awkward or controversial. The essential elements of a Commission of Inquiry are prescribed in enabling legislation: specifically (i) the terms of reference for the scope of the Inquiry are set by Government, (ii) statutory procedures are prescribed for notice and comment (submissions by the public) and (iii) discretion for public hearings where oral evidence could be examined. An interim report setting out an analysis of all submissions received, findings of fact and draft recommendations is released for further public comment. Following review of submissions received, the final report and recommendations are prepared and submitted to Government.

A Commission of Inquiry acts as a fact-finding agent for the Government, rather than as a court of law. Any conclusions or decisions made are not binding on Government. The most likely area of concern arises in respect of the decision by Government following the Commission of Inquiry, as the final decision is a political one. The Helsham Commission of Inquiry into World Heritage Listing of Tasmania's forests is an example where the recommendations of the Inquiry were virtually ignored by the Federal Government. Subsequently, in a public statement made to a national forestry conference at Launceston, the head of this Inquiry, Mr Justice Helsham, was reported in the print media — the *Hobart Mercury* (27 March 1990) — as saying:

> "Such commissions were not much use in resolving resource exploitation problems … such inquiries and commissions were not set up to resolve conflict or

assist in decision-making, but to defer decision-making so that politicians could defer the matter until it had lost its urgency or could otherwise be neutralised."

Over the last decade, the approach taken by Government to involve the community in environmental conflicts is to rely on the provisions contained in contemporary environmental legislation for consultation and participation rather than on a Commission of Inquiry. Involvement of the community is often prescribed as a means for achieving the object (or purpose) of the statute. For example, achieving the statutory object of the Federal statute, the *Environment Protection and Biodiversity Conservation Act 1999 (Cth)* (the "EPBC Act") requires a partnership approach to environmental protection and biodiversity conservation to be promoted, e.g. through the involvement of the community in management planning: Section 3(2)(g)(iv).

Environmental impact assessment is undertaken under the EPBC Act. However, there are no prescribed procedures for involving the community set out in the EPBC Act or Regulation; nor is there an express policy. Rather, reliance is placed on notice and comment provisions prescribed under the EPBC Act (Department of the Environment and Water Resources, 2007a). Contemporary Federal environmental legislation for biotechnology, the *Gene Technology Act 2000 (Cth),* also prescribes requirements for notice and comment for specific administrative decisions relating to the release of genetically modified organisms in the environment.

(b) The United Kingdom

A significant direction for public participation and procedural environmental rights was the signing by the UK of the Aarhus Convention on 25 June 1998. The UK Government then ratified the Aarhus Convention on 23 February 2005. Article 1 of the Convention refers to "*the right of every person*", *inter alia*, "*to live in an environment adequate to his or her health and well-being*". This aspiration was acknowledged as being fully shared by the United Kingdom at the time of ratification. Moreover, in ratifying the Aarhus Convention the UK gave an undertaking to guarantee, under Article 1, legal rights limited to the rights of access to information, public participation in decision-making and access to justice in environmental matters in accordance with the Convention (UNECE, 2006). Following the introduction of the new *Environmental Information Regulations 2004,* the UK Government considers that it is "Aarhus compliant", i.e. in terms of the procedural rights for public participation in environmental matters under the Convention (Stookes, 2005). A *Commentary* on the case, *Berkeley v. Secretary for State for the Environment, Transport and the Regions* [2001] Env. L.R. 16, reported at the end of the decision by the House of Lords, notes that for the UK, at least, public participation obligations will command increasing importance because of the Aarhus Convention and the *Human Rights Act 1998*.

Directive 2003/35/EC *providing for public participation in respect of the drawing up of certain plans and programmes relating to the environment as well as amending, with regard to public participation and access to justice* (the "Public Participation Directive") was adopted on 26 May 2003. Two other Directives, Directive 85/337/EEC *on the assessment of certain private and public projects on the environment* and Directive 96/61/EC *concerning integrated pollution prevention and control*, were so amended to ensure that they were consistent with the Aarhus Convention.

(c) The United States

A number of environmental statutes focus on participatory approaches to decision-making. They incorporate procedures for notice and comment, for example public submissions in relation to proposed actions by Federal agencies, e.g. the Environmental Protection Agency (EPA). Enabling legislation that facilitates public participation is provided for in statutes including the *National Environmental Policy Act*, the *Clean Water Act*, the *Resource Conservation and Recovery Act* and the *Safe Drinking Water Act* (Environmental Law Institute, 2001).

Furthermore, the focus of a *Public Involvement Policy*, released by the US Environmental Protection Agency in 2003, is to provide for meaningful public input and to consistently look for new ways to enhance public input. EPA staff and managers are required to seek inputs reflecting all points of view and to carefully consider this input when making decisions. In addition, the policy sets out a need for decision-making processes to be open and accessible to all interested groups. The foundation for the policy is to achieve effective public involvement for any decision or activity and is based on the following seven steps. Each step has specified recommended goals, actions and methods:

"(i) Plan and budget for public involvement activities;
(ii) Identify the interested and affected public;
(iii) Consider providing technical or financial assistance to the public to facilitate involvement;
(iv) Provide information and outreach to the public;
(v) Conduct public consultation and involvement activities;
(vi) Review and use input and provide feedback to the public;
(vii) Evaluate public involvement activities."

PARTICIPATORY APPROACHES, POWER AND ADMINISTRATIVE DECISION-MAKING

Effective public participation, according to prescribed legislative procedures or Government agency policy guidelines, represents the foundation for managing and resolving environmental conflicts. A key principle for effective public participation is that "*people need to be aware of the level of power being offered*" (Institute for Participatory Planning, 1981). For example, an approach that only provides the community with the opportunity to make submissions or recommendations to the agency as to the form the final decision should take, and which would not be binding on the agency, could be considered to place limits on the level of power offered to the community.

The three global approaches to public participation reviewed could all be seen to reflect the following situation: whilst each participatory approach emphasises fairness in information disclosure and public involvement, there is no direct community role in the actual decision-making process made by the Government agency; community inputs are essentially in the form of recommendations to the agency. The participatory elements of joint analysis and control over decisions and their implementation are limited, or even absent, thus restricting the level of power for the community to be effectively involved in terms of shared responsibility in the decision-making process.

This issue needs to be considered in the context of the existence of any unifying legal rights or principles that may have general application — or have persuasive value — for assessing the level of power available to the community for different participatory approaches in Australia, the UK and the United States.

LEGAL RIGHTS AND PARTICIPATORY APPROACHES

The following comparative review of legal rights and common law and the question of participatory approach focuses on the level of power the community may have in resolving environmental conflicts.

(a) Australia: Community Consultation

In *Leichhardt Municipal Council v Minister for Planning No. 2* (1994) 78 LGERA 146 [and on appeal to the Appeal Court of New South Wales (1995) 87 LGERA 78] the community consultation provision (Section 45) of the *Environmental Planning and Assessment Act 1979 (NSW)* was in dispute. This statutory provision specified the legal obligations for community consultation in terms of the bodies that had to be consulted by the Government agency when preparing an environmental study or draft regional environmental plan ("REP"). The following meaning for "consultation" was adopted by the court:

> "Given its ordinary or common meaning, according to the Oxford Dictionary, consultation involves the taking of counsel, seeking information or advice from another and taking it into consideration either by deliberation or in conference. **There is no imperative that the advice be accepted or that it be taken into account to any particular degree. The object of consultation is to be apprised or informed of other opinions or positions in regard to a subject before the matter for decision is finally determined**" (Author's emphasis).

Later, on appeal, the Court of Appeal of New South Wales (in a 2:1 decision) identified the legal obligations for consultation associated with this statutory provision and concluded that the Director of the Government agency was required to ensure that consultations with the bodies prescribed in the statute had been properly discharged.

> "Later events showed that the council's views had been considered in preparing the draft REP. Later events also showed that the council was unhappy with the result of the director's consideration of the council's views. To my mind, this does not amount to the consultation process not having been completed. It seems to me that it was completed, although not to the satisfaction of the council. That did not matter, so far as s 45 was concerned. **The obligation was to consult, not to agree**" (Author's emphasis).

(b) The United Kingdom: Community Consultation

There is a well-established body of administrative law in which the legal principles for community consultation have been rigorously reviewed and developed over time. Whilst not binding on courts outside the UK, nevertheless the following legal principles warrant consideration, in other jurisdictions, for their persuasive value. Consultation has been considered against the overriding need for fairness and to be part of the law of natural justice: *R (Edwards and Others) v Environment Agency and Others* [2006] EWCA Civ 877 (at paras. 90-94 and 102-106). In *R v Devon County Council, ex parte Baker* [1995] 1 All ER 73, the court stated "*the precise demands of consultation ....vary according to the circumstances ... The extent and method of consultation must depend on the circumstances. Underlying what is required must be the concept of fairness.*"

The principle for fairness in this case has been extended to a "full and fair consultation" — with an additional element for consideration: the accuracy of the information contained in the information disclosed for community comment. In *R v Secretary of State for Transport ex parte London Borough Council* [1995] EnvLR 390, the court concluded that:

> "The consultation exercise had been conducted unlawfully as the consultation document had misled the consultees as to the true basis of comparison to be applied as between 1988 and 1993 noise levels. The applicants were entitled to assume that the comparison was intended to provide a real comparison. **The applicants had a legitimate expectation that the consultation process would be full and fair**" (Author's emphasis).

A detailed analysis of the sequential requirements for a proper consultation, including the nature of the information that needed to be disclosed to enable an "intelligent response" from those consulted, was given by Lord Woolf MR, in *R v North & East Devon Health Authority, ex parte Coughlan* [2001] QB 213 (at para. 108):

> "It is common ground that, whether or not consultation of interested parties and the public is a legal requirement, if it is embarked upon it must be carried out properly. To be proper, consultation must be undertaken at a time when proposals are still at a formative stage; it must include sufficient reasons for particular proposals to allow those consulted to give intelligent consideration and an intelligent response; adequate time must be given for this purpose; and the product of consultation must be conscientiously taken into account when the ultimate decision is taken: *R v Brent London Borough Council, Ex p Gunning* [1985] 84 LGR 168."

and later, at para. 112,

> "It has to be remembered that consultation is not litigation: the consulting authority is not required to publicise every submission it receives or (absent some statutory obligation) to disclose all its advice. Its obligation is to let those who have a potential interest in the subject matter know in clear terms what the proposal is and exactly why it is under positive consideration, telling them enough (which may be a good deal) to enable them to make an intelligent response. The obligation, although it may be quite onerous, goes no further than this."

(c) The United States: Negotiated Rule-Making and Consultation

The "*Negotiated Rule-Making Act of 1990*" 5 U.S.C. [Section 556(f)] creates a consultative process for negotiated rule-making by Government agencies. The court in *USA Group Loan Services Inc. v Riley* 82 F.3d 708 (7th Cir. 1996) considered the issue whether the agency had failed to negotiate in "good faith" as it had not abided by an official's promise to abide by any consensus reached during the consultative process. The court concluded that the *Negotiated Rule-Making Act* did not make the promise enforceable, and that the practical effect of enforcing it would be to extinguish notice and comment rule-making in all cases where it was preceded by negotiated rule-making. In addition, there was no textual or other clue that the Act meant to do this "*the* [Negotiated Rule-Making] *Act does not envisage that the negotiations will end in a binding contract. The Act simply creates a consultative process in advance of the more formal arms' length procedure of notice and comment rule-making.*"

The court's reasoning in this case followed the reasoning in an earlier case: *Natural Resources Defense Council Inc. v EPA.* 859 F.2d 156, 194 (D.C. Cir. 1988).

ADR V PUBLIC PARTICIPATION AND POWER

Public participation and ADR are generally seen as two quite separate approaches for resolving environmental conflicts. The following goals of public participation have been defined by Beierle (1999):

i. Public education;
ii. Consideration of community values in the decision-making process;
iii. Trust-building with Government agencies;
iv. Reducing conflict between parties holding competing interests; and
v. Improving the substantive quality of decisions made by Government agencies.

Davies (2000) reasons that there is considerable overlap between all these goals for public participation and ADR — particularly when ADR, and negotiation, is applied to public interest environmental conflicts. However, Davies acknowledges that both processes can be distinguished in terms of their effectiveness in meeting each of the five goals; the significant distinction between ADR and public participation relates to the extent these goals are achieved. In particular, the level of power for decision-making that is available in each process.

(a) Community Consultation and Public Participation

Procedural rights, through "notice and comment" contained in environmental legislation, or a policy, which provides the community with the power to participate in resolving environmental conflicts, should not lead to any perception that the process is an "illusory bargain". Such a case might be where the outcome of participation is ultimately inconsistent with the final decision made by the Government agency. Unless there is a statutory requirement for the agency to be bound by the outcome of the consultation, there is no legal basis for the advice received during community consultation to be accepted or taken into account to any particular degree by the agency. A legal obligation to consult is not necessarily an obligation to agree — unless such power is provided for in a statute. However, such a legal obligation is generally not provided for in environmental legislation. The worst case scenario would be for a participatory approach to be seen as a process characterised as having some form of manipulation of public inputs into the agency decision-making process. In these circumstances, trust in process and the agency would decline as the process would be seen as neither transparent nor responsive to the needs of parties.

(b) ADR

In contrast to participation and consultation, ADR and interest-based negotiation enables the parties to have a direct role in finding mutually agreed solutions for a conflict by negotiation. Contemporary approaches to the use of ADR process and interest-based negotiation to negotiate agreements for environmental conflicts by consensus are based on joint fact-finding and problem-solving together with shared responsibility for decision-making (Susskind *et al.*, 2000). Decision-making in these circumstances facilitates trust-building with the Government agency as the process would be seen as transparent and responsive to the parties as power is shared to find a solution for the environmental conflict — a solution about which the parties would have a sense of ownership. Consequently, the level of power the community would have in negotiating an agreement for an environmental conflict would be significantly greater compared with the processes of consultation and participation.

(c) Implications for Resolving Environmental Conflicts

The common law principles arising from UK case law that apply to consultation when notice and comment provisions are provided in environmental legislation need to be recognised as a significant source of power for resolving environmental conflicts by negotiation. The legal principles that relate to "full and fair consultation", accuracy of the information upon which public comment is sought and the requirements for a proper consultation to ensure an "intelligent response" from those consulted, all have implications for potential judicial review challenges.

These principles all fall within the requirements for natural justice in administrative law. Where legislation provides procedural rights for notice and comment but is silent on the actual procedures for consultation, the observations of Lord Woolf MR in *R v North & East Devon Health Authority, ex parte Coughlan* are particularly significant. These legal principles are an important source of knowledge power for conflict resolution and are relevant considerations for a BATNA analysis; specifically, for evaluating whether either community consultation or public participation is the best alternative process to negotiation for resolving the conflict. A BATNA analysis may also act as a trigger for a Government agency to be willing to share its power in the administrative decision-making process and to negotiate an agreement with parties having a legally recognised interest in the conflict, or the community who have to live with the outcome of the agency's decision.

TRADITIONAL INDIGENOUS KNOWLEDGE AND PARTICIPATION

Berkes *et al.* (2001) acknowledge the potential application for the joint use of traditional knowledge and scientific knowledge for environmental management. However, there is little agreement on how this might be achieved. Rather than to integrate both forms of knowledge, one view is to keep both separate but with provisions for mutual enrichment and ongoing dialogue. Whilst this approach focuses on the importance of the cultural context of traditional knowledge, there is one qualification: unequal power relationships between the dominant society and Indigenous groups (Berkes, 1999).

The use of ADR to negotiate settlements based on joint fact-finding and problem-solving together with shared responsibility for consensual decision-making would have many advantages over public participation and consultation for resolving environmental conflicts, where traditional knowledge is an important element of the conflict — subject to the proviso that Indigenous groups have an adequate basis of power to participate effectively in conflict resolution. Where an Indigenous group has a legally recognised interest in the conflict, legal rights represent a significant source of knowledge power to facilitate the use of the collaborative problem-solving procedures of ADR to negotiate settlements.

STRATEGIC LITIGATION AGAINST PUBLIC PARTICIPATION ("SLAPP")

Many countries, such as the United States and the United Kingdom, have constitutional guarantees of free speech. In addition, in the UK, the *Human Rights Act 1998* gives "*further effect to rights and freedoms guaranteed under the European Convention on Human Rights 1950*". Article 10 of the Convention provides for the right to freedom of expression: a right that includes "*freedom to hold opinions and to receive and impart information and ideas without interference by public authority and regardless of frontiers*". Australia is an exception in not having any constitutional guarantees for free speech. Notwithstanding the existence of legal rights for free speech, a contemporary problem causing a constraint to public participation in environmental conflicts has emerged in some countries — the United States and Australia are but two examples. The existence of this constraint to public participation seems somewhat of a paradox given the legal rights to free speech in many countries.

(a) The Scope of SLAPP

A SLAPP suit is a tactic to silence those private citizens, or non-profit community groups, who oppose the activities of business and industry, particularly a development proposal subject to public controversy. The tactic is "*strategic*" in the sense it is aimed at the "ringleaders", the rationale being that if the "*ringleaders can be silenced then any opposition will also be effectively silenced*" (Kelly, 1995). In *Wilcox v Superior Court of Los Angeles County* 27 Cal. App. 4th 809 at 815–16 (1994) the Court of Appeal of California made the following observation: "*The paradigm SLAPP is a suit filed by a large land developer against environmental activists or a neighborhood association intended to chill the defendant's continued political or legal opposition to developer's plans*."

Pring and Canan (1992) have identified the following types of activities which have been the subject of SLAPP suits in the United States: "*testifying, debating, complaining, campaigning, lobbying, litigating, appealing, demonstrating and otherwise 'invoking the law' on public issues*". All the activities are protected by the US Constitution. All the activities involve an exercise of political speech — and not commercial speech of any form.

SLAPP suits are legal actions usually brought under defamation law but sometimes commercial law, e.g. consumer protection or fair trading laws. Why is liability pursued under defamation law? From a legal perspective, if for example, a developer considers that the comments made by a private citizen or community group about their proposed project or activity are defamatory, then the developer may have a reasonable expectation that they have a legal right to protect their reputation. The exercise of the legal right to pursue a defamation action against the private citizen or community group may have the consequence of constraining public participation. Because of the very large damage claims associated with SLAPP suits, they have the potential not only to intimidate, but also to deplete the financial resources of private citizens and community groups. It is clear that the fear of defending a SLAPP suit by private citizens and community groups may lead to an outcome where opposition to the development collapses — given the significant disparity in financial resources power between developers, private citizens and community groups, i.e. in effect, the opponents become "*sued into silence*" (Pring and Canan, 1992).

(b) SLAPP and the Law

The central issue is whether the operation of the democratic ideal requires that the law ensure that the community not be constrained from speaking about matters of public interest. In turn, such a principle raises the proposition whether inhibitions imposed by potential threats of defamation that seek to constrain public participation should be offset by measures available to the courts to protect the community (Levine, 2005).

Anti-SLAPP Legislation

Walters (2003) considers that the only solution to protect the community from SLAPP suits, and the intimidation associated with defamation actions to silence opposition, is through the introduction of specific legislation. Many jurisdictions in the United States have now introduced legislation making SLAPP suits unlawful and to encourage public participation. However, there are no examples of anti-SLAPP legislation in Australia today.

A good example of an anti-SLAPP statute in the United States is the *California Code of Civil Procedure* section 425.16(a) first passed in California in 1992. The legislative history, as well as the statute, notes that "*it is in the public interest to encourage continued participation in matters of public significance, and that this participation should not be chilled through abuse of the judicial process.*" The foundations for the statute to apply are (i) a cause of action (ii) arising from (iii) any act (iv) in furtherance of (v) the person's right of petition or free speech (Arkin, 2003). Based on an analysis of case law and legislative change over time for California's anti-SLAPP statute, Arkin (2003) concludes that the statute:

> " … provides an exemplary case-study for the proposition that judicial policymaking will rein in abuse of a statutory or common law scheme, even before legislative action is possible. Here … several appellate courts have engaged in the placement of appropriate limitations on the statute in order to return it to its original focus of protecting true political speech and petitioning activity."

Common Law Defence

The tort, collateral abuse of process, has been the subject of analysis by Kelly (1995) as an option for a defendant to a SLAPP suit in Australia. For a defendant to succeed on such a plea, it must be shown that the plaintiff is using the process of the court to achieve a purpose that is not within the legitimate scope of the proceedings. That purpose need not be the only purpose for which the plaintiff is pursuing the action, but must be the predominant purpose. Accordingly, the purpose for which the plaintiff institutes proceedings is the central issue. However, the tort raises a number of problematic areas. The first question becomes when is a purpose outside the legitimate scope of the process of the court and how to characterise the purpose? This difficult question was raised in the case of *Goldsmith v Sperrings* [1977] 1 WLR 478 where the English Court of Appeal said that if a litigant had a genuine cause of action which the litigant would wish to pursue in any event, then it would be doubtful whether the tort could be made out, notwithstanding any ulterior purpose the litigant may have had.

Secondly, there is a further evidentiary issue: how to prove that someone is suing you for an ulterior purpose? What if the intention is hidden and only becomes apparent in actions which in themselves are vague? Consequently, the situation may well be that the only evidentiary basis for intention to be established would be, indirectly, by the method of inference. Finally, the practical reality of reliance on this tort, as a defence, is that this tort will not necessarily lend itself to a quick and decisive response by the legal system — notwithstanding that the very nature of SLAPP suits warrants such a response (Kelly, 1995). Therefore, one inevitable conclusion arises. In jurisdictions where SLAPP suits may be a problem, the legislative pathway represents the preferred pathway for offsetting "SLAPP constraints" relative to the use of this tort, either as a weapon in the defendant's armoury or as a means of redress.

SLAPP: IMPLICATIONS FOR PUBLIC PARTICIPATION

In a review of Australian SLAPP cases, Walters (2003) cautions community groups involved in environmental conflicts over development proposals about describing specific legal actions as SLAPP suits. An Australian example is cited where it was claimed by a community conservation group that legal mechanisms were used by the developer to silence the community group from public participation. A defamation action succeeded in relation to the following statement published in the journal of a State conservation body: "*We believe that this legal process is being used simply to silence us ... The mechanism being used here is one called a 'SLAPP suit' commonly used in the United States to silence environmental groups, consumer groups and legitimate viewpoints being put from the community ...*" . Walters advises prudence, in this regard, as the ultimate issue as to whether a defamation suit might be regarded as a SLAPP suit is a matter for the courts to determine.

Finally, consider the following scenario. As part of the notice and comment procedures for the environmental impact assessment ("EIA") process for a development proposal, public advertisements sought public submissions in relation to the material published on environmental impacts contained in the Draft EIA. In such a situation, what might be the reaction of the proponent to a public submission made by a community group that was highly critical of the scientific methodology and conclusions by the environmental consultants who prepared the Draft EIA? Clearly, this issue may represent an area of uncertainty for public participation in the EIA process in the absence of an anti-SLAPP statute.

FREEDOM OF INFORMATION LEGISLATION AND PUBLIC PARTICIPATION

The ideals of openness and accountability by Government can be traced back to the ancient Greeks (Australian Law Reform Commission, 1994). As Riley (1987) points out, "*since the development of the system of democratic government, there has always existed a tension between the elected and the electorate as to how much the former should tell the latter*". The first legislated regime of FOI began in Sweden in 1776 (Australian Law Reform Commission, 1994). FOI legislation, the *Freedom of Information Act*, 5 U.S.C., was introduced in the United States in 1966 and then followed in other Western countries. All 50 States now have State laws that are reasonably consistent with the Federal Act (Doty, 2000). In Australia, Federal FOI legislation, the *Freedom of Information Act 1982 (Cth)*, modelled on the US legislation, was introduced in 1982 where it became the first national FOI legislation introduced into a country with a Westminster-style government. There is FOI legislation in all Australian States and Territories. The FOI legislation throughout Australia is broadly similar in scope but can be distinguished in the manner by which this is achieved (Paterson, 2005). FOI legislation, the *Freedom of Information Act 2000*, was introduced in the United Kingdom in 2000.

(a) The Purpose of FOI Legislation

The principal objective of FOI legislation is to enhance the principles upon which democratic government is based so that the public have a broad, but a legally enforceable, right of access to official information held by Government agencies. The generally accepted basis for FOI legislation includes:

- To give the public a legally enforceable right of access to information held by Government agencies;
- To facilitate public scrutiny of actions by Government and, in turn, to promote accountability of government; and
- To promote public participation in government.

(b) FOI Legislation as a Source of Power for Public Participation

By having access to information held by Government agencies, through FOI legislation, citizens become better informed and have an increased awareness of potential environmental and human health problems arising from administrative decisions made by Government agencies. Past examples from the United States of information associated with environmental conflicts that has been obtained under FOI legislation (Wald, 1984; Barkas, 2005), include:

- Information on radioactive and toxic drinking water in New Mexico;
- Information on high rates of cancer amongst plutonium workers;
- Information on high birth defects in Utah due to atomic bomb testing; and
- Information on harmful drugs and serious health hazards.

Legal rights provided under FOI legislation to access documents or information held by Government agencies are a significant source of knowledge power for the community to participate effectively in conflict resolution. The power dynamic in public interest environmental conflicts between Government and the community reflects the extent documents or information may be restricted from disclosure by FOI legislation. The level of power may be initially skewed against the community when information is not freely available, but may be offset if documents and information become available under FOI legislation. In addition, FOI legislation facilitates the identification of any documents or reports which may be subsequently sought, should the conflict be the subject of data mediation where ADR is used to manage the conflict.

(c) FOI Legislation as a Constraint to Public Participation: Exempt Documents

A person seeking documents or information belonging to a Government agency under FOI legislation is not necessarily entitled to any of the materials that are sought. Documents or information may not be disclosed where they come within the category of "exempt" under FOI legislation. Exemptions under FOI legislation need to be understood as they may limit the rights of a citizen or a community group to have access to documents or information held by Government. Therefore, an exemption may represent a constraint to limit the power to effectively participate in conflict resolution. Some general examples of the exempt category include a document that, if disclosed, would adversely affect the public interest, e.g. documents relating to national security. In addition, documents that would damage the business affairs of an agency if disclosed, e.g. documents such as trade secrets that require confidentiality to be maintained, would generally come within the exempt category. The exemption provisions are intended to achieve a balance between the public interest in providing access to Government information and the need to protect other legitimate interests that may be affected by disclosure (Paterson, 2005). The extent public participation and Government accountability can be promoted by access to information through FOI legislation is directly related to the range of categories of documents or information that are exempt.

It should be understood that FOI legislation does not prevent the disclosure of an exempt document or information. Rather, there is no positive duty to provide access to a document that is exempt (Paterson, 2005). The burden of proof for disclosure falls on the Government agency to which the request has been made and not on the person who is seeking the information.

LEGISLATIVE APPROACHES TO THE REGULATORY CONTROL OF INFORMATION

FOI legislation enables the community to have access to official "*documents*" or "*information*" held by Government agencies. Australian FOI legislation provides for access to documents. In contrast, FOI legislation in the United Kingdom and the USA provides for access to information. There is an important distinction between the terms "documents" and "information" that needs to be recognised, as the form of material that may be disclosed is dependent on the meaning of these terms.

(a) Australia

The Federal FOI legislation is far more restrictive as to the material that may be disclosed, relative to the UK and the United States. The *Freedom of Information Act 1982 (Cth)* has exemption provisions for protecting (i) governmental and public interests and (ii) the internal processes and the business affairs of Government agencies. Paterson (2005) provides the following analysis of the scope of exemptions under the Federal FOI statute:

- Documents that would be exempt under the category of "*governmental and public interests*" include (i) documents affecting Commonwealth–State or interstate relations, and (ii) Cabinet/Executive Council documents — including documents submitted or proposed to be submitted to Cabinet, and documents disclosing decisions and deliberations. Documents containing purely factual material are not exempt documents provided they do not disclose any deliberation or decision of Cabinet or the Executive Council.
- Documents relating to internal processes of Government agencies are exempt in specific circumstances. The document must come within the description of a "*deliberative process*". In addition, the document must relate to decision-making as part of the "deliberative processes" of the agency, or Minister or of the Government. Such documents will be exempt if their release would disclose records of opinion, consultation or deliberation that takes place within this context.

Paterson (2005) considers that the exemption of "deliberative processes" has the potential to hide from scrutiny the true basis upon which Governments make decisions and so act as a constraint to effective public participation. Concern has also been raised at "*blanket classes of exemptions*" which are frequently relied on in refusing FOI applications. These include Cabinet-in-Confidence, Commercial-in-Confidence, "internal working documents" and FOI requests that would involve an "*unreasonable diversion of resources*" (Herman and Ryan, 2004).

Furthermore, Paterson (2005) states that the fact that the FOI legislation refers to documents rather than to information represents a defect in the legislation. Paterson contends that confining access to information in documents is inconsistent with modern data-handling, e.g. the accumulation of information in databases rather than documents *per se*. In addition, it may also be conducive to the deliberate non-recording of information as a mechanism for maintaining secrecy.

(b) The United Kingdom

Section 74 of the *Freedom of Information Act 2000* provides for regulations to be made for the purpose of implementing the information provisions of the Aarhus Convention. The provisions of the Convention are similar to, but not the same as, the provisions of the FOI Act. When regulations are made pursuant to Section 39 of the FOI Act (a provision dealing with "*exempt information*"), disclosure of environmental information is made under the *Environmental Information Regulations 2004* and not under the FOI Act. Environmental information is therefore not exempt information (McDonald and Jones, 2003). Under Article 2.3 of the Aarhus Convention, environmental information means "*any information in written, visual, aural, electronic or any other material form*" and includes:

- The state of elements of the environment [paragraph (a)].
- Factors such as substances, energy, noise and radiation [paragraph (b)].
- The state of human health and safety [paragraph (c)].

The *Environmental Information Regulations 2004* prescribe exceptions rather than exemptions (as is the case under the FOI Act), for disclosure under an FOI search. The specific exceptions under the *Environmental Information Regulations 2004* have been reviewed and analysed by Coppel (2005). All of the following exceptions are qualified with the statement that "*subject to the public interest test, a public authority may refuse to disclose information to the extent*":

i. That the request relates to material which remains to be completed, to unfinished documents or to incomplete data. This exception also extends to disclosure of "internal communications" where such information is held by an agency;
ii. Disclosure would adversely affect international relations, defence, national security and public safety;
iii. Disclosure would interfere with the course of justice;
iv. Disclosure relating to confidentiality or to third party information being adversely affected;
v. Disclosure would adversely affect intellectual property rights; and
vi. Disclosure would adversely affect the protection of the environment to which the information relates.

The final exception relates to information that includes personal data; no "public interest test" applies for this exception.

Coppel (2005) concludes that the exceptions under the Regulations are generally narrower than the exemptions under the FOI Act. Furthermore, given the fact that most of the exceptions require the public interest test to be applied, Coppel concludes that it would be difficult to foresee many cases in which a request for disclosure might be refused under the Regulations — which is not necessarily the case for the FOI Act.

(c) The United States

The *Freedom of Information Act* 5 U.S.C. Section 552 ("FOIA") was passed in 1966 and has been subject to amendment from that time. The FOIA requires agencies to make information available to the public by the following means (Pierce, 2002):

- A requirement for agencies to publish specified types of information in the Federal Register: section 552(a)(i); or
- A requirement for agencies to make available for public inspection and copying (i) all final opinions in adjudications, (ii) all statements of policy or interpretation that are so narrow and particularised that they need not be published in the Federal Register, as well as (iii) staff manuals and staff instructions that affect members of the public: section 552(a)(ii); or
- Section 552(a)(iii) prescribes that except for the records that are made available under Sections 552(a)(i),(ii) where a request is made for records which reasonably describes the records and is made in accordance with published rules and procedures that are to be followed, the agency shall make the records promptly available to "*any person*".
- Under the FOIA, the term "information" is given a very wide interpretation and includes records, files, data, reports and maps. A copy of a document requested under the FOIA must be provided unless an agency has good reason to withhold all or part of a document because disclosure is considered to be inappropriate.

Section 552(b) prescribes nine circumstances where information is routinely exempt from disclosure. These exemptions need to be considered in the context of the following legal principle: that exemptions under the FOIA "*do not obscure the basic policy that disclosure, not secrecy, is the dominant objective of the* [FOI] *Act*" and that the exemptions "*must be narrowly construed*": *Department of the Air Force v Rose* 425 U.S. 352 (1976):

i. Classified national security and foreign relations matters;
ii. Matters related to an agency's internal personnel rules and practices;
iii. Trade secrets and privileged commercial or financial information;
iv. Inter-agency or intra-agency memos or letters that would be considered privileged in litigation;
v. Personal privacy files;
vi. Records kept for law enforcement purposes;
vii. Oil and gas-well data and maps;
viii. Information dealing with regulation or supervision of a financial institution; and
ix. Information specifically exempted from disclosure by another statute where the statute provides no discretion other than to withhold the matter from the public.

In relation to the last category of exemption (withheld from disclosure by statute), it is relevant that many environmental statutes contain specific provisions to make certain information or data available to the public (Environmental Law Institute, 2001). These include:

i. *National Environmental Policy Act* ["NEPA"] 42 U.S.C. Section 102(2)(G): information useful in restoring, maintaining and enhancing the quality of the environment;
ii. *Emergency Planning and Community Right-to-Know Act* 42 U.S.C. 11001 *et seq.* (1986) enacted by Congress as the national legislation on community safety: provides public information to help protect public health, safety and the environment from chemical hazards;
iii. *Clean Air Act* ["CAA"] 7 U.S.C. Section 114(c): availability of records, reports or information pertaining to compliance certificates and monitoring data — subject to the exception that the information is protected as a trade secret; and
iv. *Comprehensive Environmental Response, Compensation and Liability Act* ["CERCLA"] 42 U.S.C. Section 117(b): all final remedial action plans are made available before remedial actions are commenced.

FOI: IMPLICATIONS FOR PUBLIC PARTICIPATION

Accessibility of information held by Government is directly related to the legal rights contained in FOI statutes which permit disclosure. There is a very broad spectrum evident in the legislative approaches in the legal right the public have to access Government documents and information — from the formidable statutory provisions constraining disclosure of documents in Australia to the more enlightened pathway facilitating disclosure of environmental information in the United Kingdom.

Australia

The potential constraints to effective public participation in Australia are significant since the view that legislation intended to make information accessible to the public has, in practical reality, resulted in information being blocked (Herman and Ryan, 2004). These authors do raise the scenario that it does not necessarily follow, where a document falls within an exempt category, that the information must be withheld. That is, a Government agency has discretion whether or not to disclose exempt documents. However, their conclusion is that this discretion is not exercised in favour of greater openness of Government. In addition, based on an analysis of a substantial body of evidence — reports of the Ombudsman, the Australian Law Reform Commission, Parliamentary Committees, published articles in the literature and statistics contained in the reports of the operations of FOI legislation — Paterson (2005) concludes that FOI statutes, throughout Australia, "*have fallen a long way short of achieving their objective to extend as far as possible the community's right of access to government information*".

The United Kingdom

In the United Kingdom, it is significant that the framework for FOI legislation is linked to two fundamental principles of the Aarhus Convention — access to environmental information and public participation in environmental decision-making. The later development of EU Directive 2003/4/EC *on public access to environmental information* is particularly significant as a complement to the Aarhus Convention as it recognises that the disclosure of official information to "*the widest extent possible*" needs to be considered along with effective public participation in environmental decision-making.

For the United Kingdom, Coppel (2005) states that "New Year's Day 2005 should be proclaimed as the day on which there will be a revolution in the United Kingdom" — the day that rights that enliven the *Freedom of Information Act 2000* took effect. A significant step in the FOI legislative framework for the UK is that specific requests for environmental information come within a special statutory framework, as provided for in the *Environmental Information Regulations 2004.* It is Coppel's view that the significant step in FOI legislation arises because the new direction for general rights of access to information is bound to shift the balance of power in planning and environmental law given the recognition for a wider access to official environmental information.

The United States

The position in the United States in terms of public access to environmental information held by Government is one in which availability is facilitated by the FOIA as well as a number of environmental statutes. Despite claims that the exemptions are sometimes arbitrarily enforced, Doty (2000) states that the FOIA has been "*relatively successful*" in being able to make Government information open to citizens. Doty identifies the most serious obstacle as "*bureaucratic inertia in the agencies themselves*", particularly when an FOI request involves controversial information. However, on 14 March 2007, *FOIA Amendments of 2007* were passed by the House having the purpose of increasing public access to Government information by offsetting long delays and bureaucratic obstacles experienced by those requesting information.

It is also relevant that, in the United States, the FOIA is supplemented by what are called "sunshine laws". Sunshine laws provide rights of access to meetings of major decision-making bodies and not just to records of them — unless closed in accordance with any of the nine exceptions of the FOIA. Under Section 552(b) of the *Government in the Sunshine Act* ("GSA"): "*Members shall not jointly conduct or dispose of agency business other than in accordance with this section. Except as provided in subsection (c)* ['in a case where the agency finds that the public interest requires otherwise'] *every portion of every meeting of an agency shall be open to public observation.*"

4. Enforcement of environmental laws: legal rights, conflict resolution, knowledge power and negotiation

"Avoiding the battlefield mode of dispute resolution is a growing and desirable trend, and probably 90% of legal controversies are ultimately settled out of court. The problem is that, without the portent of a legal battle, it is highly unlikely that market players will willingly sit down to negotiate any compromise … Realistically, many environmental cases must turn to the force of law to give effect to their arguments — the threat of court action to force negotiated consideration of public interest environmental values or the actuality of legal action when threats are not enough."

Plater *et al.* (2004)

"Hundreds of environmental agreements are negotiated every week in the United States. Many involve the settlement of pending litigation … The agreements reached often fall short … of addressing the underlying concerns that brought the parties together in the first place. Most represent little more than one side or the other 'giving up', at least temporarily … The effectiveness of settlement agreements depends on how well the parties … are able to deal with their differences through negotiation. Unfortunately, they are not often able to do it very well."

Susskind *et al.* (2000)

INTRODUCTION

The eminent jurist, Lord Denning (1979) made some incisive observations on the intrusive nature of Government in relation to how it has concerned itself with almost every aspect of life in the contemporary world. For example, Government agencies have been given much statutory power, which may be exercised with unfettered discretion in many directions such as regulating housing, employment, planning and a host of other activities. A statutory power enables Government to do what otherwise would be illegal subject to the power being exercised within legal limits. The danger to the ordinary person, Lord Denning reasons, lies in the fact that all power is capable of misuse or abuse. There would be little dispute that the observations by Lord Denning clearly extend to the increasing power Government has, today, in relation to the regulatory control of environmental protection and management. Administrative decisions made by Government agencies impact on developers as well as the community who have to live with the outcome. The potential for the abuse or misuse of statutory power associated with the intrusive nature of Government reflects the need to have the means for keeping Governments in check and accountable for their administrative decision-making (Das, 2001). For environmental management and protection issues, the relevant matters that must be taken into account by Government agencies as a legal obligation in their administrative decision-making process are prescribed in environmental legislation. Consequently, an understanding of the appropriate principles in public law — the branch of law governing the relationship between private individuals and Government — is a pre-condition to keep Government agencies accountable. Public law is a source of knowledge power to address any misuse or abuse of statutory power in the administrative decision-making process by a Government agency.

Knowledge power, through an understanding of the legal rights that parties may have to pursue some form of a civil enforcement action, is the essential starting point. Next, is an awareness of how this knowledge may be applied to facilitate the use of ADR and negotiation to resolve an environmental conflict. The key legal process reviewed is judicial review of administrative decision or actions made by Government agencies under environmental legislation. Some consideration will also be given to enforcement actions by way of a merits review, as civil enforcement actions may be undertaken as either a judicial review or a merits review.

Civil enforcement actions are also governed by the legal rights a party has to access a court to be heard on the issues in dispute. Private citizens and third parties, such as community organisations, may have restrictions in this regard as access to a court is not necessarily equal for everyone. There are well-established rules for "standing to sue" (or *locus standi*) which determine the rights of access to a court in any civil enforcement action. The legal rights for standing are an integral part of undertaking a civil enforcement action. If the rules for standing cannot be met, then the opportunity to pursue a civil enforcement action is problematic.

LEGAL RIGHTS AND KNOWLEDGE POWER: JUDICIAL REVIEW OF ADMINISTRATIVE DECISIONS

Judicial review is a legal process that provides courts with the capacity to provide remedies to people who have been adversely affected by a particular administrative decision made by a Government agency. It may involve unlawful Government action or inaction, or the limits on power conferred by law being exceeded through a gross excess of authority or an abuse of power, e.g. where something is done for the wrong reason, or by the wrong procedure, and cannot be justified (Wade and Forsyth, 2004). Judicial review may only be used in public law matters. For example, a person, whether a citizen or not, may challenge an administrative decision made by a Government agency that may have the potential to cause, or has actually caused, adverse impacts on the environment.

The basis of a judicial review action is to give effect to the following recognised public law values (Administrative Review Council, 2006):

- Safeguarding individual rights of individuals to bring a legal action to enforce a right or to prevent unlawful conduct by a Government agency;
- Ensuring the accountability of Government and its agencies to properly exercise their powers in the administrative decision-making process; and
- Ensuring consistency and certainty in the administration of legislation.

Most administrative decisions made by Government agencies may be subject to judicial review unless a statute expressly precludes judicial review or the action is subject to agency discretion by law (Ferrey, 2004; Administrative Review Council, 2006). Whether an administrative decision made by a Government agency will be considered to be an unlawful Government action will depend on a number of factors, such as:

i. The source and nature of the power that is being exercised under the specific environmental statute; and
ii. The nature of the administrative decision.

APPROACHES TO JUDICIAL REVIEW

Whilst there is some overlap between Australia, the UK and the United States for the various grounds for a judicial review challenge, there are some quite specific differences. All three jurisdictions have grounds based on illegality — such as an error of law or a failure to observe procedures required by law ("procedural impropriety") — as well on grounds based on an unreasonable exercise of power.

(a) Procedural Fairness and Procedural Due Process

Procedural impropriety is a ground for review. In Australia and the UK, the basic concept is now referred to as "procedural fairness" (formerly "natural justice"). In the United States, the elements for procedural fairness that relate to just and fair processes in legal proceedings are reflected in the comparable concept of "procedural due process". The Fifth Amendment of the United States Constitution provides that no person shall "*be deprived of life, liberty, or property, without due process*". Procedural due process implies official action must meet minimum standards of fairness to the individual and the right to a meaningful opportunity to be heard (Gellhorn and Levin, 1997). For the United States, in relation to public participation and "due process", Parkowski and Teichman (2004) contend, by reference to the Constitution, that for a party to succeed in a claim that the quantity and quality of the participatory process afforded them were inconsistent with due process they would have to show that the Government agency had, at some time, deprived the party as members of the public generally, of any liberty or property interest without due process of law.

Of special importance in the position for public law in Australia and the UK, is that the elements of procedural fairness have been extended to incorporate a principle of "legitimate (or reasonable) expectation": the judgement of Lord Denning MR in *Schmidt v Secretary of State for Home Affairs* [1969] 2 Ch 149 and by the High Court of Australia in *Kioa v West* (1985) 159 CLR 550. What is of particular significance for judicial review and environmental conflicts are the common law legal principles developed by UK courts in relation to procedural fairness, legitimate expectation and information disclosure and public participation. For the United States, the development of the common law for due process has recognised legitimate expectation as a concept but in more limiting circumstances, notably for issues such as employment and privacy. For example, the Court of Appeal, in *Livsey v Salt Lake County*, 275 F.3d 952, 957-8 (10th Cir. 2001), 112 made the following observation:

> "The Supreme Court has recognized that one component of the right to privacy is 'the individual interest in avoiding disclosure of personal matters …'
>
> We have further explained that: Due process … implies an assurance of confidentiality with respect to certain forms of personal information possessed by the state. Information falls within the ambit of constitutional protection when an individual has a legitimate expectation … that it will remain confidential while in the state's possession … "

(b) Merits Review

Another major difference between the three jurisdictions is that judicial review in the United States also extends to what in Australia (and the UK) would be understood as a merits review[3] of agency policy decisions (Sackville, 2000). In contrast, the position in Australia is that courts exercising powers of judicial review must not intrude into the merits of administrative decision-making. As is the case for Australia, there is no consideration of the merits of a case in a judicial review in the UK. Judicial review in the UK is not, and has never been seen as, an appropriate procedure for resolving disputed questions of fact (Beloff, 2005).

GROUNDS FOR REVIEW: AUSTRALIA, THE UK AND THE UNITED STATES

The *Administrative Decisions (Judicial Review) Act 1977 (Cth)* (the "AD(JR) Act") is the key Federal statute for providing the Federal Court of Australia with the jurisdiction to undertake judicial review of administrative decisions made by Federal Government agencies under Commonwealth legislation, or by a Federal agency under a State or Territory statute. There is no judicial review statute in the UK, as in Australia and the United States. English public law, with minor exceptions, is entirely judge-made. The *Civil Procedure Rules* facilitate rather than "*constrain or constrict the judicial process*" (Lord Justice Sedley, 2004). In the United States, judicial review of administrative actions of Federal agencies is provided for under Chapter 7 of the *Administrative Procedure Act* ("APA") 5 U.S.C. Section 702: "*A person suffering legal wrong because of agency action, or adversely affected or aggrieved by agency action within the meaning of a relevant statute, is entitled to judicial review thereof.*"

(a) Australia and the UK

The grounds for judicial review in Australia are prescribed in Section 5(1) of the AD(JR) Act. These statutory grounds are comparable with the common law grounds developed in public law for the UK.

Grounds for Review

The general grounds for review in Australia and the UK have been summarised by Das (2001):

[3] A merits review requires the substantial factual questions in issue or grounds of the claim to be taken into account in deciding a case.

- Illegality, e.g. an error of law in the administrative decision-making process or a decision that was not authorised by the statute under which it was made or procedures that were required by law to be observed in connection with the making of the decision were not observed;
- Irrationality (or unreasonable or improper exercise or excessive use of power), e.g. consideration of an irrelevant matter in the administrative decision-making process or where there is no evidence or other material to justify the making of the decision;
- Procedural impropriety, e.g. a breach of natural justice; and
- The principle of proportionality.

Illegality or error of law Das (2001) and the Administrative Review Council (2006) consider that the first ground, illegality or error or law, represents a formidable check and balance for the abuse of power by an agency. It requires the administrative decision-maker to satisfy the court that the decision was authorised by the law. This ground for review ensures that an administrative decision-maker acts within their authority and correctly applies the law that governs the power that may be exercised. This would include interpreting the law correctly and exercising the power only for a statutory purpose.

Irrationality The second ground for judicial review, irrationality, arises where there has been no illegality in the administrative decision that has been made. Rather, the decision is unreasonable. An unreasonable decision would be one in which the decision-maker had failed to take relevant factors into account — or alternatively, had taken irrelevant factors into account in arriving at the decision made (Das, 2001; Administrative Review Council, 2006). In *Greenpeace Ltd. v Secretary of State for the Environment, Food and Rural Affairs* [2005] EWCA Civ 1656, the Court of Appeal made the following conclusion in relation to the relevant matters considered in the administrative decision-making process:

> "The point taken is … an alleged failure to have regard to a material consideration … [The] decision-maker is generally the master of the matters he will treat as relevant and the depth or detail into which he will go: see … *CREEDNZ v. Governor-General* [1981] 1 NZLR 172 [a statute identifies expressly or impliedly considerations that must be taken into account by the court as a legal obligation] applied in English law by the House of Lords in Re Findlay [1985] AC318 … "

The simplest case is where the legislation exhaustively defines the matters that must be considered in the administrative decision-making process. In these circumstances, any other matter will be irrelevant. Common examples of relevant matters contained in environmental legislation are the specific criteria prescribed for granting a licence or a permit; or the factors prescribed for determining the critical habitat of a threatened species. The only relevant matters that the administrative decision-maker can consider are those prescribed in the statute. The matters may all be binding on the administrative decision-maker as relevant matters, e.g. "A recovery plan must consider factor X, factor Y and factor Z". The statute therefore defines the relevant matters to be considered. If a matter other than one prescribed in the statute is considered, then the question of irrelevancy arises.

Procedural impropriety Procedural impropriety requires the administrative decision-maker to adhere to the well-established rules of natural justice (now more commonly referred to as "procedural fairness") towards persons who will be affected by the decision. There are three elements to procedural fairness:

- A right for an individual affected by the decision to be heard — unless that right has been clearly excluded by statute;
- The administrative decision-maker to be free of bias towards any person affected by the decision; and
- The principle of legitimate (or reasonable) expectation for the right to be heard.

Proportionality The principle of proportionality requires that the administrative measures that are taken should not be more severe than is necessary for achieving the desired result. Proportionality requires a court to consider whether actions or measures taken were reasonable and necessary and did not exceed the limits of what was necessary to obtain the legitimate objectives of the legislation in question. As Lord Justice Sedley (2004) observed, the proposition that an administrative decision made by Government, to be lawful, must be proportionate, will have little meaning unless the answer to the question can be evaluated against some standard. Rather than to rely on conclusions that the administrative decision was "perfectly reasonable" or "manifestly absurd", Lord Justice Sedley advances the following standards to assess proportionality:

- Does the measure not merely limit, but destroy the right?
- Have relevant and sufficient reasons for adopting it been shown?
- Was there a less restrictive alternative?
- Has the choice or measure been arrived at fairly?

(b) The United States

The broad grounds for judicial review are prescribed in the *Administrative Procedure Act* ("APA"), 5 U.S.C. Section 706(2)(B) and relate to errors of law, procedural impropriety and, in finding facts or exercising discretion, whether the decision made by the agency was reasonable:

> "(A) arbitrary, capricious, an abuse of discretion, or otherwise not in accordance with law;
> (B) contrary to constitutional right, power, privilege, or immunity;
> (C) in excess of statutory jurisdiction, authority, or limitations, or short of statutory right;
> (D) without observance of procedure required by law;
> (E) unsupported by substantial evidence … ; or
> (F) unwarranted by the facts … "

Asimow *et al.* (1998) and Gellhorn and Levin (1997) provide the following commentary for these grounds. An agency's findings of basic fact are central for any judicial review challenge as to the "*substantial evidence*" and "*clearly erroneous*" tests. In making an administrative decision, an agency must interpret and apply the meaning of words, including scientific terms and concepts, according to their statutory definitions. Following the findings of basic facts and the interpretation of the governing law, the agency must apply that interpretation to the facts. Discretion for the agency in its administrative decision-making process is provided under the APA under the "arbitrary and capricious" test. This test is applied to consider whether the action by the agency was "*arbitrary and capricious, an abuse of discretion or otherwise not in accordance with law*" on the basis that the agency correctly applied the law and adopted a rational view of the facts. In reviewing an abuse of discretion, the question may also arise whether the agency had genuinely engaged in reasoned decision-making and had taken a "hard look" at the issues. The "hard look" doctrine refers to the detailed and intensive inquiry undertaken by courts in their review of administrative discretion.

JUDICIAL REVIEW AND ENVIRONMENTAL CONFLICTS

Information conflicts are a primary source of conflict for sustainability and the environment and so warrant close scrutiny in any potential judicial review challenge. There is a significant distinction between Australia and the UK with the United States and the scope of a judicial review in addressing facts. Information disclosure for public comment, as a ground for review, is also a key determinant for assessing the outcome of any prospective challenge in the UK and, in all likelihood, Australia.

(a) Australia

In addition to the Federal AD(JR) Act, each Australian State has enacted judicial review legislation to undertake judicial review of administrative decisions made by a State Government agency under State legislation. Sackville (2000) states that there are two accepted pillars for judicial review of administrative decisions that are fundamental in determining the proper scope and set the boundaries for judicial review in Australia:

- "That courts exercising powers of judicial review must not intrude into the 'merits' [i.e. the substantial question in issue or grounds of the claim] of administrative decision-making … "; and
- "That it is for the courts and not the executive to interpret and to apply the law … "

Accordingly, the situation in Australia, for judicial review, may be summarised as follows (Administrative Review Council, 2006):

i. It is the role of the court, not the Executive, to make binding determinations on the meaning and application of the law. The court's jurisdiction does not go beyond the declaration and enforcement of the law;
ii. The administrative decision-maker has complete authority to make determinations on the facts of the case, subject to any qualifications contained in the judicial review statute, e.g. the "no evidence" rule; and
iii. The "principle of deference" to the Executive on questions of statutory construction that has been followed in the United States (*Chevron's* case), does not apply in Australia.

The no evidence rule and facts

The AD(JR) Act provides for no evidence as a ground for review, i.e. whether there is any evidence to support the findings of fact made by the administrative decision-maker: ss. 5(1)(h). In addition, this ground can arise indirectly under grounds such as procedural fairness, unreasonableness, taking into account irrelevant considerations and an error of law (Morgan and Airo-Farulla, 1997). In their analysis of case law, Morgan and Airo-Farulla conclude, notwithstanding that most findings of fact will not be reviewable decisions under the AD(JR) Act, that the ground of no evidence can be made out if it comes within either of the following situations:

- The party challenging the decision establishes that the fact critical to making the decision did not exist. The fact in question need not be the sole or even the predominant critical factor but must be found by the administrative decision-maker; or
- It can be shown there is an absence of material such that the conclusion of fact made by the administrative decision-maker was not reasonable.

(b) United Kingdom

There are two dimensions where UK public law has developed principles for addressing the issue of facts and judicial approach.

Procedural fairness and legitimate expectation: public notice and comment

A judicial review challenge based on the grounds of procedural fairness, specifically legitimate expectation, may arise in environmental conflicts through the community consultation process. People who are consulted are entitled to a legitimate expectation that the consultation process will be full and fair. This is a well-established common law principle in UK public law: see *R v Secretary of State for Transport ex parte London Borough Council* [1995] EnvLR 390 and *R v Falmouth & Anor ex parte South West Water Ltd* [2001] QB 445.

A person asserting that a legitimate expectation exists must be able to point to something in the nature of a policy, promise, practice or other conduct by the agency which gives rise to it. In the UK case *R (Bibi) v Newham LBC* [2002] 1 WLR 237, Lord Justice Schiemann extracted the following principle from the common law:

> "In all legitimate expectation cases, whether substantial or procedural, three practical questions arise. The first question is to what has the public authority, whether by practice or promise, committed itself; the second is whether the authority has acted or proposes to act unlawfully in relation to its commitment; the third is what the court should do."

The issues in the UK case, *R (on the Application of Greenpeace Ltd) v Secretary of State for Trade and Industry* [2007] EWHC 311 (Admin) related to public consultation procedures following the release of an Energy White Paper setting out the Government's proposals for a future nuclear energy policy. The court considered whether there had been a breach of the claimant's legitimate expectation to "*the fullest public consultation*" — a requirement of the Energy White Paper. In finding that the consultation process was procedurally unfair and that there had been a breach of the claimant's legitimate expectation that there would be "the fullest public consultation", Sullivan J decided that "*the decision in the Energy Review that new nuclear build 'has a role to play ... ' was unlawful*" and concluded:

- " ... *it is difficult to see how a promise of anything less than 'the fullest public consultation' would have been consistent with the Government's obligations under the Aarhus Convention*": para. 51.
- " ... *the overriding requirement that any consultation must be fair is not in doubt. What is fair, and in particular whether fairness demands that new material which has not been available during the consultation period should be made available to consultees so that they have an opportunity to deal with it before a decision is taken, must depend upon the particular circumstances of the case*": para. 61.
- "*The consultation exercise was very seriously flawed*". Information on the issue of waste contained in the Consultation Document was described by Sullivan J as wholly inadequate and misleading. Information, on an issue of such importance and complexity that was given in a later consultation paper outlining proposals was described as wholly insufficient to enable the consultees to make "*an intelligent response*": para. 116.
- " ... *all, or virtually all, the information of any substance* [on both the economics and waste issues] ... *emerged after the consultation had concluded*". Moreover, "*there could be no proper consultation, let alone 'the fullest public consultation' as promised in the 2003 White Paper, if the substance of these two issues was not consulted upon before a decision was made. There was therefore procedural unfairness, and a breach of the claimant's legitimate expectation that there would be 'the fullest public consultation' before a decision was taken to support new nuclear build*": para. 117.

These public law principles developed in the UK for legitimate expectation and information disclosure as part of the community consultation process would not be binding in Australia, but would have persuasive value, given that legitimate expectation is a long-held accepted principle of the law of procedural fairness in Australia.

No evidence rule and error of material fact

A court will quash an erroneous and decisive fact that was either:

- Jurisdictional;
- Found on the basis of no evidence; or
- Wrong, misunderstood or ignored, i.e. an error of material fact.

Wade and Forsyth (2004) state that as very few findings of fact are jurisdictional, a potential judicial review challenge therefore needs to be able to identify what findings of fact arising under the no evidence rule or as an error of material fact, could come within the grounds for a judicial review application.

The no evidence rule No evidence does not simply mean a complete lack of evidence to support a finding, but extends to a case where any conclusion to support the finding, based on the evidence as a whole, would be unreasonable. A finding that cannot be supported, in this regard, is an abuse of power as it may cause a grave injustice and so comes within the principle of "ultra vires"; that is, beyond the scope of power allowed. Wade and Forsyth (2004) state that the no evidence rule has some similarity to the hard look doctrine in the United States where the court closely scrutinises whether an agency decision was based on a consideration of all relevant factors.

Error of material fact Legal principles reviewed in English case law by Wade and Forsyth (2004) define error of material fact as a ground for judicial review in terms of (i) misunderstanding, (ii) ignorance of an established and relevant fact or (iii) acting upon an incorrect basis of fact. The authors refer to the following observations made by Lord Wilberforce in *Secretary of State for Education and Science v Tameside MBC* [1997] AC 1014 (at 1030) to clearly illustrate the legal principles that apply for an agency to be exercising its powers reasonably:

> "If a judgement requires, before it can be made, the existence of some facts, then, although the evaluation of those facts is for the Secretary of State alone, the court must inquire whether those facts exist, and have been taken into account, whether the judgement has been upon a proper self-direction as to those facts, whether the judgement has not been made upon other facts which ought not to have been taken into account. If those requirements are not met, then the exercise of judgement, however bona fide it may be, becomes capable of challenge."

Judicial review and European Union law

Given the dominance of EU law in the UK, it needs to be recognised that actions (and inactions) of EU Authorities are also subject to judicial review by the Court of Justice (see Article 230, para. 4 of the *EC Treaty of Rome*). The grounds for judicial review are reasonably analogous to the common law principles established in the UK. However, the rules for standing may have limitations compared with the wider rules for standing in UK domestic law.

(c) United States

Plater *et al.* (2004) point out that judicial review challenges to actions by agencies relating to the environment may be made over a continuum of difficulty: varying from a basic application and interpretation of a statute (e.g. whether there has been an error of law in making the decision) to the more difficult assertion that the exercise of power by the agency may have resulted in a clear error of judgement. The grounds for judicial review are prescribed in Section 706 of the APA. Plater *et al.* (2004) state that most substantive challenges to administrative decisions made by Federal agencies relate to challenges on the merits. Specifically, whether the administrative decision can be substantiated in relation to (i) the arbitrary, capricious or abuse of discretion test [*APA* Section 706(2)(A)] and (ii) the requirement for substantial evidence [*APA* Section 706(2)(E)]. However, the authors acknowledge that there may be some subjectivity as the review turns on the question of whether the administrative decision made by the agency was reasonable in the circumstances.

The Supreme Court of the United States has adopted a totally different approach from the scope of judicial review compared with Australia and the UK. As Sackville J (2000) points out, the Supreme Court has followed a principle of "deference to the executive" on questions of statutory construction. Moreover, through what has been termed the "hard look" doctrine, it has simultaneously been prepared to carefully study the merits of agency decisions.

The principle of deference

The leading case on the principle of deference to the executive on questions of statutory construction is *Chevron USA, Inc. v Natural Resources Defence Council, Inc.* 467 US 837 (1984) — a case that has been applied to administrative decision-making by agencies having regulatory control for environmental protection. The *Chevron* decision established the following two-step approach to questions of statutory construction (Sackville J, 2000):

- *Step 1:* Where the intent of Congress is clear as to "*the precise question at issue*", the reviewing court and the agency must give effect to the expressed intent of Congress.
- *Step 2:* If however, the reviewing court determines that Congress has not directly addressed "the precise question in issue", the court does not simply impose its own construction of the statute. Rather, if the statute is silent or ambiguous with respect to the specific issue, the question for the court is to determine whether the agency's answer is based on a permissible construction of the statute.

In the later case of *United States v Mead Corp.* 533 U.S. 218 (2001), the Supreme Court concluded that the rule of deference to agency interpretation did not arise where the agency made a determination that was not embodied in a regulation, but only applied to an interpretation where it appeared that the agency had exercised its authority delegated by Congress to make rules carrying the force of law. Accordingly, Ferrey (2004) has summarised the position on deference as follows:

- General judicial deference to the substance of administrative rules applies where there are disputed issues of technical facts or policy or where the statute does not precisely answer the question the rule addresses (as in the *Chevron case*); and
- Where the reviewing court categorises the disputed issue as a question of law, courts tend to decide the question independently, rather than to defer to the determination by the agency.

The hard look doctrine

Sackville J (2000) considers that this doctrine is an "offset" to the *Chevron* principle. The effect of the hard look doctrine is that the Federal Court scrutinises agency decisions very thoroughly and in depth, in what would be considered to be a merits review in Australia and the UK. The key provision, APA Section 706 (2)(A), requires the reviewing court to "*hold unlawful and set aside agency action, findings, and conclusions found to be arbitrary, capricious, an abuse of discretion, or otherwise not in accordance with the law*".

Sackville J (2000) refers to the legal principles arising in *Citizens to Protect Overton Park, Inc. v Volpe* 401 U.S. 402 (1971) as a landmark case for the hard look doctrine. The court's role was to consider whether there had been a clear error of judgement in the decision to authorise the use of public funds to finance the construction of an interstate highway through a public park — and whether all relevant factors had been evaluated. The Supreme Court held that the Secretary of Transportation's decision would be upheld if the reviewing court was able to find that the Secretary could have "*reasonably believed*" that there were no feasible alternatives to the proposed construction of the highway. In particular, whether all relevant factors had been considered in making the administrative decision and whether there had been a clear error of judgement.

An outcome of the hard look doctrine is that Government agencies require detailed records to explain the basis of decisions they have made as well as ensuring they have carefully evaluated the available factual evidence (Sackville J, 2000). Doremus (2005) states that whilst there are no reports of the "Daubert standard" being applied in the judicial review of Government agency decisions, at least one Federal Appeal Court (in *Niam v Ashcroft* 354 F.3d 352 (7th Cir. 2004)) made an observation that "junk science" should be kept out of agency decisions: "*Junk science has no more place in administrative proceedings than in judicial ones*".

LEGAL RIGHTS AND KNOWLEDGE POWER: PERFORMANCE OF PUBLIC DUTIES AND JUDICIAL REVIEW

An illegal action may also arise in administrative decision-making when a Government agency fails to perform a duty. A duty may be distinguished from a power because of the prescriptive nature of a duty compared with the discretionary nature of a power (Harding, 1989). A public duty is simply a duty of a public body arising in public law. The primary source of such public duties is legislation. Given the upsurge in contemporary environmental legislation, a prudent step would be to review the legislation to identify any public duties which may impose a legal obligation upon Government agencies. In the United States, the citizen suit provision of the CWA (*Clean Water Act* 33 U.S.C. Section 1365) is a typical example of similar provisions in many environmental statutes; it authorises citizens to commence a civil enforcement action where there has been an alleged failure to perform a mandatory duty, i.e. a non-discretionary duty (Miller, 2000). In addition, international environmental conventions may create statutory schemes for liability for damage caused by global environmental hazards. For example, oil pollution or a nuclear accident may impose legal duties for the signatory party where the convention has become part of the domestic law (Stanton *et al.*, 2003).

(a) Identifying and Enforcing a Public Duty

A case for a public duty to arise may be through a statutory command, in what have been described as mandatory phrases, such as "*It shall be the duty of A to do Y*". In addition, a duty may be implied even when there is no such obligatory wording, e.g. "*an entitlement of B to have Y done by A*" (Harding, 1989). However, not every statutory provision in legislation creates a public duty so that a statutory duty may, or may not, be legally enforceable.

The approach of courts to determine the question of whether a statutory duty is to be enforced is to require that the duty be sufficiently specific, rather than being too vague or of a general or undefined character (Wade and Forsyth, 2004). Harding (1989) gives specific examples of imperfect duties that may be unenforceable such as (i) general directives, (ii) duties that relate to the provision of services, (iii) duties that are significantly qualified or contain large elements of discretion and (iv) duties expressed to be enforceable.

(b) Remedies to Enforce the Performance of a Public Duty

The remedy for a breach of statutory duty in a private law action is compensation for damages. Such a remedy can be distinguished from a public law action where, in this case, the duty is owed to the public generally and not to a private citizen. The remedy for actions in public law is to enforce the performance of a public duty imposed by statute (Jones, 2006). In the UK, there have been a number of decisions of the House of Lords involving local authorities, statutory duties and the general public interest, e.g. *O'Rourke v Camden L.B.C.* [1998] AC 188. In *X (Minors) v Bedfordshire CC* [1995] 2 AC 633, the House of Lords decided that a failure by local authorities in their statutory duties did not give rise to a claim for damages in private law. These decisions have led Jones (2006) to conclude that actions in public law to enforce the performance of a statutory duty are to be brought by way of judicial review. In addition, Wade and Forsyth (2004) conclude that the most important remedy to enforce the performance of a statutory duty by public authorities is mandamus: a writ compelling a Government official or agency to correctly perform mandatory duties. Miller (2000) also refers to the use of mandamus actions in the United States, where public law actions have been brought under citizen suit statutes. The position in Australia seems to be that the enforcement of public duties is presently reviewable by judicial review, but through the remedies of declaration or injunction (Schoombee, 1998).

REMEDIES FOR JUDICIAL REVIEW

The power of the court to give remedies following a judicial review is limited to upholding, quashing or declaring administrative decisions as invalid for being made unlawfully (Beloff, 2005). The available remedies at common law enhance the options open to the community (Administrative Review Council, 2006; Wade and Forsyth, 2004) and include:

Writ of mandamus: Compels the person to whom it is directed to perform a public duty that it had failed to perform. The ground for mandamus is that the decision, or conduct, under review is subject to legal error and would render performance of the duty ineffective.

Writ of prohibition; order for certiorari: Prohibition and certiorari are closely related remedies. Prohibition forbids the party to whom the writ is directed from doing something which would be beyond power, e.g. something that is proposed to be done. The ground for prohibition is that the decision, or conduct, under review is subject to legal error and so unlawful. Certiorari quashes a decision which is beyond power. Certiorari looks to the past, prohibition to the future.

Injunction: An injunction enforces statutory obligations upon Government or others. An injunction is more flexible than the other remedies and allows the person, subject to the injunction, an opportunity to rectify the problem.

Declaration (or "declaratory order"): A declaration is made by a court and is a statement of the law or a statement of the rights and duties of a party, following, for example, an administrative decision made by a Government agency. A declaration results in a court order being made that applies to the specific situation. In Australia and the UK, the grounds for a declaration may relate to the agency decision and procedural fairness, whether a public duty exists or whether an administrative action may be unlawful because of an error of law; it has no restraining effect as a court can only state what the law is as it has no jurisdiction to remedy the error in the administrative decision (Amos and Churches, 1998). However, the court order can be a useful remedy against a Government agency because it could be expected to give effect to the court order without any compulsion (Administrative Review Council, 2006; Tromans, 2006). In the United States, an agency under APA Section 554(e), in its discretion, "*may issue a declaratory order*" in order "*to terminate a controversy or to remove uncertainty*". The order is binding on both the Government agency and the party seeking the declaration.

For environmental conflicts, a declaration may be a broad and useful remedy for determining whether an administrative decision or action may be unlawful because of an error of law. Contemporary environmental legislation contains many scientific terms and concepts as part of the subject matter and purpose of the statute. For example, terms and concepts such as "biological integrity", "critical habitat", "ecological health", "environmental harm", "resilience" and "sustainable development" are common questions of law contained in legislation. Therefore, a declaration may be a remedy to pursue in order to determine whether a Government agency, in its administrative decision-making process, has correctly applied the legal meaning of a scientific term or concept as prescribed in the legislation. In addition, where the legislation fails to define a scientific term or concept, there may be uncertainty between the agency interpretation of the term or concept and its accepted scientific usage and interpretation.

F(utility) of Judicial Review Remedies

Although the remedies of judicial review are discretionary, Beloff (2005) reasons that this, in itself, is an encouragement for the courts to grant them as courts can balance the public and private interests involved in an orderly and systematic way. Moreover, Beloff emphasises that a judicial review challenge will not result in the court overturning an administrative decision if the decision does not misapply the law or is not procedurally flawed. In judicial terms, such an administrative decision is impregnable and will not be overturned merely because the court disliked the decision. Significant factors identified by Caldwell (1986) upon which the exercise of discretion is dependent include:

- Whether appropriate, alternative remedies were available but had not been used;
- The nature of the administrative decision under review: the effects of a breach on the outcome of the decision;
- Whether there was unwarranted delay in seeking relief on the part of the aggrieved party, or conduct contributed to the defect complained of; and
- Whether the relief granted was futile, having no practical value or consequence.

However, Beloff (2005) does raise a cautionary note as to the potential for a successful judicial review challenge to be futile. Where a judicial review challenge succeeds, there is always a risk that, following referral of the administrative decision to and proper reconsideration by the administrative decision-maker, the same outcome as the original decision may re-emerge. The counter-argument to this proposition is that such an outcome does not necessarily follow if a successful judicial review challenge succeeded on the grounds of a failure to provide a full and fair disclosure of information for public comment; or on the grounds of the "no evidence rule/unreasonableness" (for Australia and the UK) or the "hard look doctrine" (for the United States). In these circumstances, the utility of a successful judicial review challenge is demonstrated as the reconsidered decision would be made by the administrative decision-maker with information that would not have been available at the time of the original decision.

OPPORTUNITIES FOR JUDICIAL REVIEW

Civil enforcement actions through a judicial review for violations, or threatened breaches, of environmental laws related to the regulatory control of environmental management and protection will invariably involve an administrative decision made by a Government agency to grant a licence, permit or authority — with or without conditions. A judicial review challenge could be made in Australia, the UK and the United States by a body or industry, the subject of regulatory control under environmental legislation. By having a legal right to seek a permit or licence for an activity associated with environmental impacts, the body or industry would have a legally recognised interest under the statute. In addition, judicial review challenges may be made by members of the public and other third parties, such as community or environmental groups, in relation to administrative decisions made by Government agencies — provided statutory or common law standing requirements are met.

Tromans (2006) considers that, for the UK at least, community action and reliance on judicial review "*is one of the most fertile areas of litigation which has grown phenomenally in recent years*". Gellhorn and Levin (1997) describe a similar situation for the United States: " ... *the trend of the last two decades has been breaking down ... barriers and allowing liberal access to judicial review. The threshold defenses still have some bite, but review has become freely available to an extent that would have surprised lawyers of a generation ago.*"

LEGAL RIGHTS AND KNOWLEDGE POWER: MERITS REVIEW AND ENFORCEMENT OF ENVIRONMENTAL LAWS

Civil enforcement actions for actual or threatened breaches of environmental legislation may also be undertaken as a merits review. Criminal enforcement actions of environmental laws may also be undertaken as a merits review.

The Scope of a Merits Review

A merits review, as part of a civil or criminal enforcement action, can be distinguished from a judicial review. A merits review is directed at the substantial questions in issue, or the grounds of the claim that are to be taken into account in deciding the case. Resolving the substantial questions in issue in environmental litigation through a merits review will be dependent on the evaluation of scientific and technical evidence and expert opinion. The relevant questions of law prescribed in the statute set the boundaries for the actual scientific evidence. Divergent expert scientific opinion on disputed complex and numerous facts will have to be decided by the court. The court then makes findings of fact on the disputed facts, applies these findings to the relevant questions of law and then makes its decision. As is the case for judicial review, the legal requirements for standing may restrict members of the public and third parties who have a right to undertake a merits review to enforce an environmental law by way of a civil action.

Criminal enforcement actions are the exclusive province of Government regulatory agencies, notwithstanding that some statutes, e.g. in Australia and the United Kingdom, may allow citizens to bring private prosecutions in specific situations.

LEGAL RIGHTS AND KNOWLEDGE POWER: STANDING TO SUE AND ADMINISTRATIVE DECISION-MAKING

Not everybody who is involved in a civil enforcement action — either a judicial review or a merits review — is automatically entitled to proceed to a court hearing for a judicial determination of their case to have a legal right or duty enforced. Courts, in either private or public law, will only hear cases where the parties have some legally recognised interest in the subject matter of the dispute. Standing to sue (or *locus standi*) is a legal concept that must be satisfied for a party to be entitled to have access to the court. It is regarded by the courts, particularly in public interest actions, as a threshhold issue to be resolved.

In Australia, the United Kingdom and the USA, a legal right for standing may be granted by legislation (*a statutory right*) or may arise under the common, or general, law (*a common law right*). The statute or common law principle specifies the nature of the legally recognised interest in the subject matter of the dispute that must be satisfied for standing to sue. The Constitutions of some countries may also provide a basis for access to justice and standing to sue. For example, Article 88 of the 1991 Constitution of Colombia states that anyone who has a collective right can sue to protect it. In addition, other Constitutions state forms of actions (such as *amparo* in Peru) that have been interpreted to allow *acciones populares*, i.e. popular actions by any citizen (Bonine, 1998).

(a) Australia: Standing to Sue and Judicial Review

Judicial review applications to the Federal Court under the *Administrative Appeals (Judicial Review) Act 1977 (Cth)* [the "AD(JR) Act"] requires two central tests to be satisfied for standing. First, the person making the application must be a "*person who is aggrieved by*" a decision, conduct or failure to make a decision by the Government agency: Sections 5, 6 and 7. Additionally, an "*aggrieved person*" must come within the statutory meaning for this term [at AD(JR) Act s. 3(4)] by having "*interests ... adversely affected*" by the decision, conduct or the failure to make a decision. The following two cases illustrate the contemporary approach taken by the Federal Court for standing and public interest environmental disputes.

Australian Conservation Foundation v Minister for Resources (1989) 19 ALD 70 involved the peak national conservation body in Australia in a judicial review challenge. A judicial review of an administrative decision made to grant a woodchip export licence was sought under the AD(JR) Act, on the basis that the decision may adversely affect the natural heritage values of native forests listed under the under the *Australian Heritage Commission Act 1975 (Cth)*. Davies J acknowledged that the tests for standing under Subsections 3 and 5 of the AD(JR) Act were wide and flexible. Whilst the Australian Conservation Foundation did not have standing to challenge any decision which might affect the environment, Davies J concluded it had a special interest in the subject matter of the dispute — the South East Forests. Some of the forest areas were on the "Register of the National Estate", a Register of sites of national significance for their natural and/or cultural heritage values. The following conclusions in the decision of Davies J illustrate the essential elements that led to a special interest being satisfied:

> "(a) The present issue was not a local issue.
> (b) The public perception of the need for the protection and conservation of the natural environment and the need for bodies such as the [Australian Conservation Foundation] to act in the public interest had noticeably increased in the last decade.
> (c) The [Australian Conservation Foundation] was not just a busybody in the area but was established and functions with governmental financial support to concern itself with such an issue and is pre-eminently the body concerned with that issue."

Davies J made the following observations in this case:

> "In determining standing, it is necessary to take account of current community perceptions and values"

and later:

> "In my opinion, the community at the present time expect that there will be a body such as the [Australian Conservation Foundation] to concern itself with this particular issue and expects the Australian Conservation Foundation to act in the

public interest to put forward a conservation viewpoint as a counter to the viewpoint of economic exploitation".

In a later case heard by the Federal Court under the AD(JR) Act, the reasoning of Davies J granting standing to the peak national conservation body in *Australian Conservation Foundation v Minister for Resources* was applied and extended by Sackville J to a local environmental group: *North Coast Environmental Council Inc. v Minister for Resources* (1994) 85 LGERA 270. The North Coast Environmental Council sought a declaration for the Ministerial decision to grant an export licence for the export of woodchips from forests in the region of the local environmental group. Standing under the AD(JR) Act was in issue. On reviewing case law authorities on standing, Sackville J suggested that the trend in the evolution of the test for standing for judicial review for an "*aggrieved person*", in Australia, was to allow for a wider range of people to challenge Government decisions. Because of the status of the North Coast Environmental Council and the closeness of its relationship to the subject matter of its request, Sackville J decided that the North Coast Environmental Council had a special interest to come within the standing requirements of the AD(JR) Act as a "*person aggrieved*". In addition, it had standing under the general law to challenge Ministerial decisions to grant in-principle approval for woodchip export licences. The following factors were identified in finding that the local environmental group satisfied the special interest test:

"(a) [It] was the peak environmental organisation in the north coast region of [the State] with activities related to the areas affected by the operations generating the woodchips;
(b) [It] had been recognised by the Commonwealth as a significant and reputable environmental organisation through the provision of regular financial grants;
(c) [It] had been recognised by the [State] Government as a body that should represent environmental concerns on advisory committees;
(d) [It] had conducted or coordinated projects and conferences on matters of environmental concern for which it had received significant Commonwealth funding; and
(e) [It] had made submissions on forestry management issues to [a Federal Agency]."

Standing and Federal environmental legislation

Judicial review of Federal Government administrative decisions can also be pursued under Commonwealth environmental protection legislation, e.g. the *Environment Protection and Biodiversity Conservation Act 1999 (Cth)* ["EPBC Act"]. It is particularly significant that the EPBC Act extends statutory standing for judicial review to a "*general environmental interest*" (Bates, 2006). Under s. 487 of this Act, an individual or an organisation or an association will be an "*aggrieved* person" if, at any time in the two years immediately before the decision, failure or conduct:

(i) "the individual has engaged in a series of activities in Australia for the protection or conservation of, or research into, the environment; or
(ii) The objects or purposes of the organization or association include protection or conservation of, or research into, the environment."

It is also significant that the prescribed statutory test for standing for an "*interested person*" who may apply for an injunction under the "Enforcement Part" of the EPBC Act (Part 17), has a meaning reasonably similar to the meaning for an "*aggrieved person*" seeking judicial review: see Subsection 475(6)(7).

Where standing is not prescribed as a statutory right under environmental legislation, it may be granted as a common law right if a "*special interest*" can be shown. The special interest test most commonly requires ownership of the land that is the subject matter of the dispute or, alternatively, economic benefits being received from the land.

Standing and Indigenous interests

In the absence of any statutory provisions for standing, the common law principles that specify a special interest for standing must be satisfied. The decision in *Onus v Alcoa* (1981) 36 ALR 425 is particularly significant for Indigenous groups as the High Court concluded that a "*cultural interest*" satisfied the common law requirement for a special interest. In this case, the cultural interest in the land, the subject matter of the dispute, was characterised by the fact that the Indigenous group were the descendants of the traditional inhabitants of the land and, by custom handed down from generation to generation, were the custodians of the sacred land.

(b) United Kingdom: Standing to Sue and Judicial Review

There is no judicial review statute in the UK. Applicants seeking judicial review are required to satisfy the prescribed requirement of having "*sufficient interest in the matter to which the application relates*": Section 31(3) *Supreme Court Act 1981* and Civil Procedure Rules Part 54. In their analysis of the decision of the House of Lords in *R v Inland Revenue Cmrs. ex parte National Federation of Self-Employed and Small Businesses Ltd* [1982] AC 617, Wade and Forsyth (2004) concluded that the consideration of standing is to be related to the facts of the case as it is incorrect to treat standing as a preliminary issue independently of the merits of the complaint, citing Lord Wilberforce: "*the question of sufficient interest can not, in such cases, be considered in the abstract, or as an isolated point: it must be taken together with the legal and factual context*".

In an incisive analysis of case law and legal principles applied to the interest test for standing, Beloff (2005) acknowledges that the test for standing in public law is more generous compared with private law. Although a precise test for standing is difficult, Beloff states that it is particularly relevant that the judiciary "*seeks to include rather than to exclude*" or as Tromans (2006) states, standing is now much less of a bar for judicial review challenges. Beloff (2005) gives the following examples, from judicial review litigation, of interests held by persons or organisations as examples of satisfying standing in public law:

i. Those with personal interests, e.g. parents' concern relating to electromagnetic fields from high voltage underground cables and risk to children's health;
ii. Local groups with a local interest, e.g. a planning decision by a local council;
iii. National groups with a general interest, e.g. Friends of the Earth and a potential breach of an EC Directive for Drinking Water (80/778/EEC).

The review and analysis of the judicial reasoning in *Inspectorate of Pollution ex parte Greenpeace Ltd (No. 2)* [1994] 4 All ER 329 by Beloff (2005) identifies the key factors that established Greenpeace as satisfying the sufficient interest test for standing. The court's decision reflected the following advantages Greenpeace brought to the judicial review process — notwithstanding Greenpeace's general environmental interest:

- Assisting the court to facilitate justice between the parties by a well-informed challenge;
- Having particular expertise in environmental matters including access to relevant experts in science and technology;
- Bringing issues to the court in an effective way in the form of carefully selected, focused, relevant and a well-argued challenge; and
- In addition, the court took into account that Greenpeace had participated during the consultation process.

The case, *R v Foreign Secretary ex parte World Development Movement Ltd* [1995] 1 WLR 386 related to a British consortium that sought aid and trade provisions to construct a hydro-electric power station in Malaysia. The complainants were a non-partisan pressure group dedicated to improving British aid to other countries and were not local residents. The court found a sufficient interest to justify standing by considering the following factors:

- "*The importance of vindicating the rule of law*";
- "*The likely absence of any reasonable challenger*";
- "*The nature of the breach of duty against which relief is sought*"; and
- "*The prominent role of* [World Development Movement Ltd] *in giving advice, guidance and assistance with regard to aid*".

In his analysis of case law, Tromans (2006) states that judicial thinking, today, appears to centre on the approach taken by Sedley J in *R v Somerset County Council ex parte Dixon* [1998] EnvLR 111. Such an approach has clear implications for the more awkward cases for establishing standing when an "environmental interest" is in issue:

> "Public law is not about rights but about wrongs. The Courts have always been alive to the fact that a person or organization with no particular stake in the issue or outcome may, without in any sense being a meddler, wish and be well placed to call the attention of the court to an apparent misuse of power. If an arguable case for such misuse may be made out of an application for leave, the Courts only concern is to ensure that it is not being done for an ill motive."

EU administrative law, standing and judicial review

The actions and inactions by EU authorities are subject to judicial review by the European Court of Justice: Article 230, *Treaty of Rome*. However, the rules of standing are interpreted literally and give far less scope for individual complainants compared with contemporary standing rules in the UK (Wade and Forsyth, 2004). Article 230 (para. 4) of the *Treaty of Rome* limits standing to complainants having a "*direct and individual concern*", as distinct from the wider public concern (Beloff, 2005). Cygan (2003) states that granting of standing under Article 230 has been narrowly interpreted by the European Court of Justice and, in marked contrast to the trend in Member States such as the UK, standing does not extend to judicial review actions by representative groups, e.g. conservation groups. Cygan contends that the right to seek judicial review under Article 230 EC should not, in itself, open the floodgates, particularly if the Treaty was modified to allow for targeted public interest litigation to allow citizens to protect their own rights and to safeguard the rights of the community in which they live.

In ratifying the Aarhus Convention in February 2005, the UK Government gave an undertaking, under Article 1, to guarantee legal rights such as "*access to justice in environmental matters in accordance with the Convention*" (UNECE, 2006). In this regard, Article 9(3) of the Aarhus Convention may be a relevant consideration as it provides public interest groups and citizens with the standing to go to court, or another review body, to enforce an environmental law that has been contravened by the act or omission of a public authority or a private person (Brans, 2005).

(c) United States: Standing to Sue and Judicial Review

Judicial review of administrative actions of Federal Agencies is undertaken under the APA. The source for standing is Article III of the Constitution which extends judicial power of the Federal Court to "cases" and "controversies". A party seeking judicial assistance must have a sufficient stake in the dispute in order to satisfy the case or controversy requirement (Asimow *et al.*, 1998). In contrast to the more "liberal" approach to standing in the UK and Australia, judicial review of Federal agency decisions by the Supreme Court of the United States seems to have moved to a more restricted approach to citizen standing in environmental conflicts (Ferrey, 2004).

In their review and analysis of Supreme Court case law, Ferrey (2004) and Plater *et al.* (2004) characterise the requirements for standing and environmental litigation with respect to three different tests, all of which must be satisfied, for a citizen to be entitled to standing for judicial review in the Federal Court.

Test I: Whether there is "personal injury in fact" necessary to establish a Constitution Article III "case" or "controversy"

The judicial reasoning by the Supreme Court in *Sierra Club v Morton* 405 U.S. 727 (1972) in its analysis for standing concluded "*... the 'injury in fact' test requires more than an injury to a cognizable* [i.e. within the court's jurisdiction] *interest. It requires that the party seeking review be himself among the injured.*" This test continues to be contentious as it requires the citizen group to show specific injury to the organisation or its members — not injury to the environment: see *Friends of the Earth, Inc. v Laidlaw Environmental Services (TOC), Inc.* 528 U.S. 167 (2000). An interest to support a case for standing could relate to an injury to aesthetic, conservation or recreation interests, as well as economic interests (Ferrey, 2004).

Further clarity on the principles required to establish standing for the "injury in fact" test was provided by the Supreme Court in *Bennett v Spear* 520 U.S. 154 (1997). The minimum constitutional requirements imposed by Constitution Article III that must be satisfied were described as follows:

- For the injury in fact suffered, to be concrete and particularised and actual or imminent — not conjectural or hypothetical;
- For the injury to be traceable to the challenged action of the defendant, and not the result of the independent action of some third party not before the court; and
- That it be likely, not merely speculative, that the injury will be redressed by a favourable decision.

Test II: Whether the injury claimed comes within the "zone of interest" test

This test is drawn from Section 702 (Right of Review) of the APA which provides for judicial review for a person adversely affected or aggrieved by agency action. A very clear explanation of the underlying principles required to satisfy the zone of interest test was provided by the Supreme Court in *Bennett v Spear* 520 U.S. 154 (1997), as a distinction is made between the purpose of the statute and the specific statutory provision relied upon by the plaintiff:

> "In determining whether the petitioners have standing under the zone of interests test to bring their [Administrative Procedure Act] claims, we look ... to the substantive provisions of the [Endangered Species Act], the alleged violations of which serve as the [essence] of the complaint ... **Whether a plaintiff's interest is 'arguably ... protected ... by the statute' within the meaning of the zone of interests test is to be determined not by reference to the overall purpose of the Act in question (here, species preservation), but by reference to the particular provision of law upon which the plaintiff relies** ... " (Author's emphasis).

Test III: Application of judge-made prudential principles to deny standing

Ferrey (2004) as well as Plater *et al.* (2004) consider that principles relevant for any determination of standing that apply to this test would include:

i. Whether a final administrative decision had been made by the agency, i.e. the action was "*ripe*" for judicial review. Section 704 (Actions Reviewable) of the *Administrative Procedure Act* provides that only "final" agency actions may be subject to judicial review; and
ii. Whether all possible administrative remedies had been demonstrated to be exhausted prior to seeking judicial review. In *McKart v United States* 395 U.S. 185 (1969), the Supreme Court, in its reasoning, referred to the doctrine of exhaustion of administrative remedies: "*that no one is entitled to judicial relief for a supposed or threatened injury until the prescribed administrative remedy has been exhausted*". Applying the doctrine of exhaustion to the APA means that if there is another cause of action against the agency that could be pursued, it must be used before a judicial review could be commenced.

(d) Australia: Civil Enforcement Actions, Standing to Sue and Merits Review

The Commonwealth Administrative Appeals Tribunal [the "AAT"] has the jurisdiction to undertake merits review of Ministerial and Commonwealth Government agency administrative decisions under prescribed provisions for over 400 Commonwealth statutes. Not all these statutes deal with environmental management and protection. The environmental statutes that are most frequently the subject of merits review before the AAT are the *Great Barrier Reef Marine Park Act 1975 (Cth)*, the *Environment Protection and Biodiversity Conservation Act 1999 (Cth)* and the *Fisheries Management Act 1991 (Cth)*. A merits review before the AAT in relation to a Commonwealth statute is governed by the wide standing provisions of the *Administrative Appeals Tribunal Act 1975 (Cth)*. An application may be made by or on behalf of "*any person whose interests are affected by the decision*": ss. 27(1). Moreover, standing is also given to interest groups. An organisation or association (whether incorporated or not) will be taken to have interests that are affected by the decision, if the decision relates to a matter included in its objects or purposes — subject to the proviso that the decision was made either after the formation of the organisation or association, or the objects or purposes included the matter concerned: ss. 27(2) (3). Many Australian States also have administrative appeals tribunals, e.g. Victoria, New South Wales and Western Australia, which undertake merits review of Government decisions made under State legislation–as well as planning courts and tribunals that review Government decisions made under planning legislation.

There is a great diversity in legislation for environmental management and protection in each State and so considerable scope for a merits review for the appropriate case. Statutory schemes generally define the rules for standing more clearly, relative to the common law requirements, and may also extend the persons who would be granted standing (Bates, 2006). Bates states that contemporary trends throughout Australia reveal that the requirements for statutory standing in environmental legislation vary over a continuum from (i) an expansive statutory standing making merits review available to "*any person*" e.g. *Environmental Planning and Assessment Act 1979 (NSW)* to (ii) clearly defined standing criteria that relate to environmental interests and the public interest, e.g. *Environmental Protection Act 1994 (Qld)*, through to (iii) the absence of any rules for statutory standing. In the latter situation, common law principles that specify a special interest for standing must be satisfied.

(e) The United States: Standing, "Citizen Suit Statutes" and Civil Enforcement Actions

The most direct way, in the United States, for achieving standing for a judicial review — or a merits review — may be through a "citizen suit statute". Citizen suits are provided for in a number of Federal environmental statutes, as well as by many State legislatures (May, 2004). Actions brought under citizen suit statutes have been categorised by Miller (2000) as:

i. Actions for judicial review brought under the *Administrative Procedure Act* to challenge agency decisions as contrary to statutory authority;
ii. Mandamus actions against the Federal Government under citizen suit statutes common to Federal environmental statutes; and
iii. Actions brought under citizen suit statutes to enforce violations of environmental statutes by those being regulated by the agency, e.g. citizen suits against polluters who are in breach of regulatory requirements. In this regard, Campbell (2006) notes that once standing has been established in the citizen suit, plaintiffs have been successful on the merits, in a number of circuits.

An example of a Federal statute providing standing through a citizen suit is the *Endangered Species Act* ["ESA"], 16 U.S.C. Section 1540 (g):

> "A person may commence a civil action on his own behalf … against any governmental instrumentality or agency who is alleged to be in violation of any provision of this chapter [35, Endangered Species] or regulation issued … [or] against the Secretary where there is alleged a failure by the Secretary to perform any act or duty under s. 1533 [Determination of endangered and threatened species] which is not discretionary with the Secretary."

The citizen suit provision for standing in the ESA, as is the case for all Federal citizen suit statutes, contains several qualifications. For example, it is qualified in relation to the time an action may be commenced with respect (i) to the time after written notice of the reasons why an emergency is thought to exist with respect to the endangered or threatened species in the State concerned, or (ii) where the Secretary has commenced and is diligently prosecuting action to determine whether such an emergency exists. However, where there is "*alleged a failure by the Secretary to perform any act or duty under s. 1533 which is not discretionary with the Secretary*", an action may be brought immediately following notification where an emergency poses a significant risk to the well-being of any species of fish or wildlife or plants. Other examples of Federal statutes that provide citizen suits include 15 U.S.C. Section 2619 ToSCA (*Toxic Substances Control Act*), 33 U.S.C. Section 1365 CWA (*Clean Water Act*) and 42 U.S.C. Section 7604 CAA (*Clean Air Act*).

Many, but not all, of the States have citizen suit statutes that relate to environmental management and protection (Defenders of Wildlife and Center for Wildlife Law, 2000). May (2004) highlights the fact that civil enforcement under many State environmental citizen suits provides citizens with the authority to enforce State environmental laws generally; other citizen suit statutes allow citizens to sue agencies to perform mandatory duties. May considers that an example of a most powerful citizen suit statute is the *Michigan Environmental Protection Act* [*Michigan Compiled Laws,* Section 324.1701–1706 (2003)]. This Act allows citizens to sue to enforce all State laws governing the environment. Other States having a reasonably analogous citizen suit statute to Michigan include South Dakota, New Jersey, Minnesota, Massachusetts and Louisiana (May, 2004).

It needs to be recognised that a citizen suit statute does not necessarily provide an aggrieved party with standing to pursue judicial review in a Federal Court. The statutory jurisdiction of the Federal Court to hear and decide cases is subject to the limits of the Constitution, e.g. Article III. Standing is dependent on satisfying the three different tests discussed earlier. However, there is a distinction between Federal legislation and citizen suit statutes in State legislation for satisfying the requirements for standing. Based on their analysis of case law, the Defenders of Wildlife and Center for Wildlife Law (2000) conclude that, whereas the Federal Constitution imposes the requirements for standing on cases that may be heard by Federal Courts, no such requirements are imposed on State courts; consequently, the standing requirements imposed by State law on cases that it may hear may be significantly less onerous relative to the Federal level. Accordingly, the authors' conclusion that State citizen suit statutes "*generally offer greater court access than citizen provisions at the federal level*" is particularly significant for citizen enforcement actions. Such a conclusion also needs to be considered in the context of the observation of Hilson (2000) that citizen suit statutes have allowed citizen enforcement to offset any limitations in the enforcement record of regulatory agencies.

Quandt *et al.* (2005) raise a cautionary note for standing under citizen suits as there may well now be a "*constitutional gatekeeper*" for a citizen to pass. In particular, the legislature may not provide certainty in conferring standing on individuals who do not satisfy the Federal Constitution requirements for standing: see *National Wildlife Federation v Cleveland Cliffs Iron Co.* 471 Mich 608; 684 NW2d 800 (2004). In the first instance that the case was heard, the administrative law judge found that the plaintiffs had no standing in the administrative matter — notwithstanding that the *Michigan Environmental Protection Act* citizen suit provision provides that "*any person may maintain an action ... for declaratory and equitable relief against any person for the protection of the air, water and other natural resources and the public trust in these resources from pollution, impairment, or destruction*": *Michigan Compiled Laws* 324.1701(1). Ultimately, the majority decision of the Michigan Supreme Court decided that the plaintiffs had established standing under the Federal Constitution — but without considering the question of whether the plaintiffs had satisfied standing under the provisions of the *Michigan Environmental Protection Act* citizen suit provision alone.

(f) Australia and the United Kingdom: Standing and Relator Actions

Where an aggrieved party does not meet the interest test for standing, a relator action is a possible approach to offset this obstacle. A relator action involves the Attorney General bringing an action to prevent some breach of the law, with the aggrieved party joined as a relator. In the Westminster System of Government, the Attorney General is the guardian of the public interest and always has standing. The discretion to decide to pursue a relator action rests with the Attorney General. However, the exercise of discretion will be confounded by political reality. The Attorney General may be reluctant to act for political reasons, if the activities, the subject of the complaint, are carried out or sanctioned by Government (Wade and Forsyth, 2004; Bates, 2006). In these circumstances, the availability of a relator action for private citizens may be little more than an illusory bargain.

Wade and Forsyth, as well as Bates, conclude that because of the more liberal and flexible approaches taken by courts for public law standing for judicial review in Australia and the UK, the role of the Attorney General in civil enforcement of environmental legislation has reduced the need for relator actions and consider that they may, in effect, be disregarded.

IMPLICATIONS FOR ADR: KNOWLEDGE POWER OF LEGAL RIGHTS, POWER RELATIONSHIPS AND ENVIRONMENTAL NEGOTIATIONS

Knowledge of legal rights, as a source of power, is central to the undertaking of a BATNA analysis in order to determine whether litigation or negotiation may be the preferred pathway for resolving an environmental conflict. The threat of an enforcement action requires a consideration of factors that focus on knowledge of legal rights. Specifically: (i) whether to pursue a civil or a criminal enforcement action, (ii) whether the standing requirements to litigate can be satisfied and (iii) whether the civil enforcement action should be a judicial review or a merits review.

(a) Civil Enforcement Actions v Criminal Enforcement Actions

Bates (2006) considers that civil enforcement actions, not criminal enforcement actions, will be the favoured method of court-based action in order to pursue those who violate environmental laws. The basis for this conclusion is the significance of the anticipatory, or preventative, characteristic of civil actions in relation to environmental harm. A prosecution through a criminal enforcement action is generally a reaction to environmental harm that has already occurred. In contrast, civil enforcement actions, such as a judicial review remedy, seek to prevent or to restrain a potential harm to the environment from occurring. Civil enforcement actions taken by individuals or community groups for actual or potential violations of environmental laws — or challenges on the legality of the administrative decision-making process — may result in an administrative order being made by the court in anticipation of environmental harm occurring, or to seek restoration or rehabilitation of the environment where harm has occurred (Bates, 2006). Furthermore, the standard of proof of causation for criminal enforcement, beyond reasonable doubt, is much higher than the standard of proof for civil enforcement, the balance of probabilities.

(b) Satisfying the Requirements for Standing to Sue

A party that cannot meet the requirements for standing will not have access to the courts to litigate their case. Consequently, where a party is considering pursuing a civil enforcement action, careful consideration would have to be given to the specific legislation under which the challenge is to be made and its complementary standing requirements. It is clearly evident that the need to satisfy standing is a paramount consideration and so may effectively limit the availability of legislation by which a civil enforcement action could be undertaken.

In the United Kingdom, there is an expansive and flexible approach to the interpretation of the common law principles for standing for judicial review. A similar situation exists in Australia for the requirements for statutory standing for judicial review and, depending on the jurisdiction — Commonwealth or State — the standing rules for a merits review. In the United States, statutory standing and common law principles as applied in the Federal Court are somewhat restrictive, so that the most prudent path to take for a judicial review or a merits review may be through a citizen suit statute of a State. The issue of forming alliances or coalitions, of groups having similar interests in the subject matter of the conflict, where at least one member group satisfies the legal standing requirements, is one means to offset restrictions imposed by standing. The concept of alliances or coalitions is discussed in Chapter 10.

(c) Judicial Review or Merits Review

A judicial review, in contrast to a merits review, is a legal process that is primarily concerned with the legality of the administrative decision-making process. Consequently, the costs for a party to undertake a judicial review will be much less than for a merits review. A merits review will incur added costs other than legal costs, because of the need to acquire scientific reports from expert witnesses in order to resolve divergent expert scientific opinion.

(d) Evaluating the Strengths and Weaknesses of the Proposed Civil Enforcement Action

The final step is to ensure that the grounds for a judicial review specific for each jurisdiction — or the expert scientific opinion evidence to address questions of law prescribed by the appropriate statute in a merits review — have been carefully reviewed and analysed and supported by the evidence to be relied upon. In particular, a party seeking a civil enforcement action, by way of a judicial review or a merits review, must ensure that the court does not see their application as vexatious or an abuse of the process of the court. Clearly, such a situation needs to be avoided by the party making the application. The following conclusions, in the Australian case, *Attorney-General v Wentworth* (1988) 14 NSWLR 48, represent a court's analysis of the factors that justify consideration whether an application may be vexatious or an abuse of the process of the court:

- A civil enforcement action being instituted with the intention of annoying or embarrassing the person or agency against whom it was brought;
- An action being instituted for collateral purposes and not for the purpose of having the court adjudicate on the issues to which they gave rise; or
- An action being untenable or manifestly groundless as to be utterly hopeless — irrespective of the motive of the party making the application.

(e) Power Relationships and Knowledge of Legal Rights

Solutions for environmental conflicts that arise through negotiation reflect the power relationships between the parties. Where power is evenly balanced, the outcome may be fair for everyone, as each party has an adequate basis of power to participate effectively. In contrast, where power is unequal, a party, such as a community or environmental group, or other third party, that cannot satisfy any requirements for standing and so has limited power in terms of legal rights, may be coerced into accepting a solution which would not be seen as a fair outcome. That is, the community group would not have an adequate basis of power to participate effectively in the conflict. In these circumstances, a solution may only be accepted because it is not the worst option. However, it has long been recognised that there are many sources of power that may be drawn upon and that it would be unusual for any party to be totally powerless (Astor and Chinkin, 1992). This is a very basic, but significant, principle to recognise for resolving environmental conflicts by negotiation. Knowledge power of legal rights related to a civil enforcement action may, in fact, be the effective trigger to provide adequate power for the community or a citizen to participate effectively with the Government agency and the proponent of the development activity in conflict resolution.

Consider a scenario where a Government agency, or a developer, did not wish to directly involve a citizen or community group in negotiations to resolve an environmental conflict. Such a situation might arise if there was a perception that the citizen or community group had no legal rights to standing to pursue a civil enforcement action. In these circumstances, even if the Government agency or the developer were prepared to directly involve the community group in negotiating a solution for the environmental conflicts, the power relationships between the parties would be markedly unequal — against the community group. Power for a community group would be maximised where standing could be satisfied under one, or more, statutes. That is, because of a threat that potential litigation could be initiated by parties that satisfy standing and where there is a risk for the agency, or the developer, that an action initiated against them would succeed.

CONCLUSION

Knowledge of legal rights for civil enforcement can be used as a means of influencing power relationships between parties to manage and resolve an environmental conflict. A prerequisite for relying on the threat of litigation as the basis for the Government agency, or the developer, to share power and to negotiate solutions for environmental conflicts with parties who have a specific interest in the subject matter of the conflict, e.g. the community, Indigenous peoples and conservation groups, is to ensure that the potential threat of a civil enforcement action is seen as real, rather than speculative or hypothetical; that there would be a reasonable likelihood if the civil enforcement action were to proceed, it would succeed. Such an assessment could be undertaken using the ADR process of early neutral evaluation and would involve all parties. Where there is a likelihood that the threat of litigation could succeed, it would act as a catalyst for the Government agency to further consider the need to share power with the parties to resolve the conflict by negotiation. At this stage, any decision by a Government agency to share power to resolve a conflict using ADR and negotiation would be facilitated by a BATNA analysis. Specifically, to assess whether a negotiated settlement, or defending a potential civil enforcement action/judicial review challenge, would be the preferred process for resolving an environmental conflict. It is quite clear that the outcome of the overall assessment is dependent on the extent to which knowledge of legal rights represents a significant source of power to provide an adequate basis for all parties to participate effectively in resolving an environmental conflict by negotiation.

5. Sustainability and the environment

"[A] fundamental bottom line has emerged: A country cannot achieve economic development when its environment is degraded, nor can it restore its environment in the absence of economic development."

Halpern (1992)

"Numerous publications have defined, redefined and scrutinized the idea [of sustainability] and applied it to most endeavours. Although its charge is undeniably critical to a quality future, and its appeal nearly universal, some commentators have argued that sustainability is conceptually too vague to have much practical value. Others have suggested that the idea has become so diluted and misapplied to be meaningless in most forums."

Kermath (2006)

"Sustainable development is impossible without transforming the legal structure within which human activities, transactions and initiatives occur. Getting the law right is an essential pre-condition for success; getting it wrong guarantees failure despite the best development efforts."

Futrell (2002)

INTRODUCTION

The concept of sustainability has long been recognised by science and is reflected in the long history of research into the management of agricultural and natural ecosystems, where long-term stability of the biological and physical resource had been the primary objective. However, following the release of *Our Common Future* by the World Commission on Environment and Development in 1987 (the "Brundtland Report") and its focus on sustainable development, the concept has significantly extended with a wider acceptance of its significance for the environment, generally. Not only has the concept of sustainable development been incorporated in an international declaration agreed to by the global community, it has also become prescribed as a statutory requirement for environmental and planning legislation in some countries, as well as being increasingly adopted as an environmental policy. From the time of the Brundtland Report, the concept of sustainable development has become a significant unifying concept for contemporary decision-making for environmental management and protection.

Science has a central role for finding solutions to conflicts over sustainable development. Not only must science be in a position to provide solutions for conflict resolution, it must also justify them in terms of explaining the relationships between ecological, economic and social considerations (Saunders, 1990). Law also plays an important, although not exclusive, role in achieving sustainability (Saunders, 1990; Dernbach, 2002a). Environmental legislation may prescribe sustainable development as an object of the statute, or as a matter that must be taken into account as a legal obligation in the administrative decision-making process by a Government agency, e.g. the granting of an authority, licence or permit under environmental legislation for a proposed action that has the potential for adverse environmental consequences. Conflicts over sustainability may also arise outside the court system, as a public interest environmental conflict where administrative decisions are made by a Government agency in relation to an environmental policy that gives effect to sustainable development. It is very clear that sustainable development has now become a key consideration for administrative decision-making by Government agencies as resolving a conflict over sustainable development is now no longer simply a scientific issue; law, policy and values may add a further dimension.

The sustainable development concept will invariably involve balancing ecological, economic, social and cultural considerations. However, the balance point between these considerations will vary according to the competing viewpoints of the parties to the conflict. Consequently, to resolve a conflict over sustainable development will necessarily require a number of options, or alternative solutions, to be considered for the specific conflict. To simply focus on evaluating a single option for sustainable development, in the likelihood that it might represent the best solution to meet the needs and concerns of all parties, is fraught with difficulty. Rather, the preferred approach would be to evaluate a range of options along a sustainable development continuum in which the balance between ecological, economic, social and cultural considerations varied.

Finally, it is important to be aware of the global significance attached to the concept of sustainable development and Indigenous people. This aspect of sustainable development is especially relevant in relation to the co-management of natural resources. Principle 22 of the *Rio Declaration on Environment and Development*, a text adopted at the United Nations Conference on Environment and Development held at Rio de Janeiro, Brazil in 1992, states:

> "Indigenous people and their communities and other local communities have a vital role in environmental management and development because of their knowledge and traditional practices. States should recognize and duly support their identity, culture and interests and enable their effective participation in the achievement of sustainable development".

THE ORIGIN OF SUSTAINABLE DEVELOPMENT AS A GLOBAL ENVIRONMENTAL CONCEPT: INTERNATIONAL DECLARATIONS

Sustainable development was defined in the Brundtland Report as:

> " ... development that meets the needs of the present without compromising the ability of future generations to meet their needs. It contains within it two concepts: the concept of 'needs', in particular the essential needs of the world's poor, to which overriding priority should be given; and the idea of limitations imposed by the state of technology and social organisation on the environment's ability to meet present and future needs."

The Brundtland Report specified the following key elements and objectives of sustainable development:

- "Reviving growth;
- Changing the quality of growth;
- Meeting essential needs for jobs, food, energy, water and sanitation;
- Ensuring a sustainable level of population;
- Conserving and enhancing the resource base;
- Reorienting technology and managing risk; and
- Merging environment and economics in decision-making."

(a) The United Nations Conference on Environment and Development "UNCED", Brazil, 1992

Following the acceptance of the Brundtland Report by the UN General Assembly, the United Nations convened a world conference at Brazil in 1992, the United Nations Conference on Environment and Development. Representatives from 172 countries sought to resolve the interrelationships between the environment and human socio-economic activities. Two important, but non-binding texts,[4] having a general scope for sustainable development, were agreed to. The *Rio Declaration on Environment and*

[4] These two texts need to be distinguished from two multi-lateral conventions adopted at Rio de Janeiro and opened for signature: the "*Convention on Biological Diversity*" and the *"UN Framework Convention on Climate Change*".

Development specified 27 universally applicable principles to facilitate international action for sustainable development. Agenda 21 provided a global future plan of action for sustainable development based, to a large extent, on the Rio Declaration principles. As Dernbach (2002a) points out, Agenda 21 was the plan of action for sustainable development; the success or failure of the Rio Declaration was dependent on Agenda 21 being implemented.

The final form sustainable development takes in each country may differ as it would be a function of that country's culture, history, economy and environment. When countries such as Australia, the United Kingdom and the United States agreed to the Rio Declaration and Agenda 21, they agreed to implement these agreements both within their national boundaries and, as well, to make them part of their foreign policy (Dernbach, 2002b).

A number of the principles of the Rio Declaration have since been included in numerous binding and non-binding international instruments since Rio, and constitute emerging customary international law rules of conduct (Kiss and Shelton, 2004). Specific examples given by the authors include:

- Public participation;
- Environmental impact assessment;
- The precautionary principle;
- Notification of emergencies; and
- Consultation for activities that have the potential to affect the environment of other States.

(b) The World Summit on Sustainable Development, Johannesburg (2002)

The World Summit on Sustainable Development held at Johannesburg, South Africa in 2002, adopted a *Declaration on Sustainable Development* confirming an intent to "assume a collective responsibility to advance and strengthen the interdependent and mutually reinforcing pillars of sustainable development — economic development, social development and environmental protection — at local, national, regional and global levels" (Kiss and Shelton, 2004).

LEGISLATIVE APPROACHES FOR THE REGULATORY CONTROL OF SUSTAINABLE DEVELOPMENT

There are significant differences between Australia, the United Kingdom and the USA in terms of the extent that sustainability has been incorporated in national legislation as a matter that must be taken into account, as a legal obligation, for administrative decision-making by Government agencies.

(a) Australia

From the time of the Rio Declaration, the concept of sustainable development has been given express legislative implementation in environmental and planning legislation by the Federal parliament, as well as by a number of Australian States. The Federal statute, the *Environmental Protection and Biodiversity Conservation Act 1999 (Cth)* (the "EPBC Act"), has one of its objects directed towards ecologically sustainable development ("ESD"): "*to promote ecologically sustainable development through the conservation and ecologically sustainable use of natural resources*": Section 3(b). In addition, sustainability is a statutory object for management plans made under another Federal statute, the *Great Barrier Reef Marine Park Act 1975 (Cth).* The purpose is "*to ensure that activities within areas of the Marine Park are managed on the basis of ecologically sustainable use*": Section 39Y(c).

A number of Australian States have also introduced environmental legislation in which ESD is a prescribed object of the legislation, e.g. the *Environmental Protection Act 1994 (Qld)* (at section 3), the *Environment Protection Act 1993 (SA)* (at Section 10) and the *Protection of the Environment Operations Act 1997 (NSW)* (at Section 3). Planning legislation in which ESD is a prescribed statutory object includes the *Environmental Planning and Assessment Act 1979 (NSW)* (at Section 5) and the *Integrated Planning Act 1997 (Qld)* (at Section 1.2.1).

Statutory definitions for ecologically sustainable development

It is particularly relevant that the statutory definitions contained in Australian environmental legislation embody many of the principles in the Rio Declaration and Agenda 21. The following examples for the statutory meaning of ESD illustrate the overlap of the legal meaning prescribed in legislation for ESD and the Rio Declaration principles, in terms of matters that must be taken into account, as a legal obligation, for administrative decision-making by Government agencies:

i. Section 3A of the Federal statute, the *EPBC Act* prescribes the following principles of ESD:

- "Decision-making processes should effectively integrate both long-term and short-term economic, environmental, social and equitable considerations.
- If there are threats of serious or irreversible environmental damage, lack of full scientific certainty should not be used as a reason for postponing measures to prevent environmental degradation.
- The principle of inter-generational equity — that the present generation should ensure that the health, diversity and productivity of the environment are maintained or enhanced for the benefit of future generations.
- The conservation of biological diversity and ecological integrity should be a fundamental consideration in decision-making.
- Improved valuation, pricing and incentive mechanisms should be promoted."

ii. Schedule 3 ("*Standard Criteria*") of the State of Queensland statute, the *Environmental Protection Act 1994 (Qld)* endorsed the "Guiding Principles for ESD" as specified in the Federal Government policy, the *National Strategy for Ecologically Sustainable Development* (Council of Australian Governments,[5] December 1992) for its definition of ESD. The National Strategy for ESD policy gave effect to both the Rio Declaration and Agenda 21.

- "Decision-making processes should effectively integrate both long- and short-term economic, environmental, social and equity considerations;
- Where there are threats of serious or irreversible environmental damage, lack of full scientific certainty should not be used as a reason for postponing measures to prevent environmental degradation;
- The global dimension of environmental impacts of actions and policies should be recognised and considered;
- The need to develop a strong, growing and diversified economy which can enhance the capacity for environmental protection should be recognised;
- The need to maintain and enhance international competitiveness in an environmentally sound manner should be recognised;
- Cost-effective and flexible policy instruments should be adopted, such as improved valuation, pricing and incentive mechanisms; and
- Decisions and actions should provide for community involvement on issues which affect them".

---

[5] The Council of Australian Governments is the peak intergovernmental forum in Australia, comprising the Prime Minister, State Premiers, Territory Chief Ministers and the President of the Australian Local Government Association.

(b) The United Kingdom

Environmental and planning legislation, introduced in the UK after the adoption of the Rio Declaration and Agenda 21 in 1992, incorporates sustainable development as either a statutory object or prescribed as matters that must be taken into account, as a legal obligation, for administrative decision-making by Government agencies. Sustainable development is provided for in both national law and European Union law.

National law and sustainable development
The *Environment Act 1995* [at Section 4(1)] has as its principal aim and objective:

> "It shall be the principal aim of the [Environmental] Agency (subject to and in accordance with the provisions of this Act or any other enactment and taking into account any likely costs) in discharging its functions so to protect or enhance the environment, taken as a whole, as to make the contribution towards attaining the objective of achieving sustainable development … ."

Planning and local authority legislation further incorporates the concept of sustainable development. Section 39 of the *Planning and Compulsory Purchase Act 2004*, at Part 3 ("Development") expressly provides for sustainable development:

> "(1) This section applies to any person or body which exercises any function –
> (a) under Part 1 in relation to a regional spatial strategy;
> (b) under Part 2 in relation to local development documents;
> (c) under Part 6 in relation to the Wales Spatial Plan or a local development plan.
> (2) **The person or body must exercise the function with the objective of contributing to the achievement of sustainable development."**
> (3) For the purposes of subsection (2) **the person or body must have regard to national policies and advice contained in guidance** issued by –
> (a) the Secretary of State for the purposes of subsection (1)(a) and (b);
> (b) the National Assembly for Wales for the purposes of subsection 1(c)" (Author's emphasis).

The *Local Government Act 2000* also prescribes sustainable development as a matter that must be taken into account as a legal obligation by the administrative decision-maker. At Section 4(1), "*Every local authority must prepare a strategy (referred to in this section as a community strategy) for promoting or improving the economic, social and environmental well-being of their area and contributing to the achievement of sustainable development in the United Kingdom*". In addition, under Section 2(1) of the *Local Government Act*, local authorities "*have the power to do anything which they consider is likely to achieve any one or more of the following objects*"– the promotion or improvement of the (a) economic well-being (b) social well-being and (c) environmental well-being, of their area.

European Union law and sustainable development

A Treaty represents the highest level of law in the European Union. The consolidated version of the *EC Treaty Establishing the European Community (2004)*, at Article 2, provides for "*sustainable development*" as the task of the Member States. This task, amongst other things, involves: "*implementing common policies or activities to promote throughout the Community, a harmonious, balanced and sustainable development of economic activities ... a high level of protection and quality of the environment ...*"

Furthermore, Article 6 of the EC Treaty gives effect to sustainable development in terms of Community policies and activities: "*environmental protection requirements must be integrated into the definition and implementation of Community policies and activities referred to in Article 3* [e.g. a policy in the sphere of the environment] *in particular with a view to promoting sustainable development*".

Statutory definitions for sustainable development

The *Environment Act 1995*, the *Local Government Act 2000* and the *Planning and Compulsory Purchase Act 2004*, do not provide any statutory definition for the meaning of "sustainable development". However, the *Planning and Compulsory Purchase Act* provides for national policies for guidance in this regard: Section 39(3) requires a person or body, exercising the function that contributes to the achievement of sustainable development, "*must have regard to national policies and advice contained in guidance*" issued by prescribed bodies.

National policy and the meaning of sustainable development

In 1999 the UK Government released a strategy for making sustainable development a reality titled *A Better Quality of Life-A Strategy for Sustainable Development in the United Kingdom*. Later, in March 2005, the UK parliament released a new and updated strategy for sustainable development titled *Securing the Future. Delivering UK Sustainable Development Strategy*. The strategy was described by the Prime Minister as "*focusing on long-term solutions, not short-term fixes*". There are four separate, but shared, strategies that provide the focus for the *Securing the Future Policy:*

- A new integrated **vision** with strong international and societal dimensions;
- A set of five **guiding principles** focusing on environmental limits that will be used to achieve the purpose of sustainable development;
- Four agreed **priorities**: sustainable consumption, climate change, natural resource protection and sustainable communities; and
- An **indicator set** that is outcome-focused with commitments to examining new indicators, such as well-being.

The set of five shared guiding principles in the *Securing the Future* policy have been agreed to by the UK Government, the Scottish Executive, the Welsh Assembly Government and the Northern Ireland Administration. The five guiding principles are central to achieving the purpose of sustainable development. The goals of the principles of *Living within environmental limits* and *A just society* are dependent on achieving the principles of *A sustainable economy, Good governance* and *Sound science.* The guiding principles are especially relevant as they incorporate the key principles stated in the Rio Declaration and Agenda 21. Consequently, the following meaning given to each guiding principle in the national policy document is particularly significant for the interpretation and application of sustainable development:

i. Living within environmental limits: "*Respecting the limits of the planet's environment, resources and biodiversity — to improve our environment and ensure that the natural resources needed for life are unimpaired and remain so for future generations.*"
ii. Ensuring a strong, healthy and just society: "*Meeting the diverse needs of all people in existing and future communities, promoting personal wellbeing, social cohesions and inclusion, and creating equal opportunity for all.*"
iii. Achieving a sustainable economy: "*Building a strong, stable and sustainable economy which provides prosperity and opportunities for all, and in which environmental and social costs fall on those who impose them (polluter pays), and efficient resource use is incentivised.*"
iv. Promoting good governance: "*Actively promoting effective, participative systems of governance in all levels of society — engaging people's creativity, energy, and diversity.*"
v. Using sound science responsibly: "*Ensuring policy is developed and implemented on the basis of strong scientific evidence, whilst taking into account scientific uncertainty (through the precautionary principle) as well as public attitudes and values.*"

EU policy and the meaning of sustainable development

The *EC Treaty Establishing the European Community (2004)* does not provide any definition for the meaning of "sustainable development". In 2000, at Lisbon, the European Council set a new strategic direction for sustainable development. In 2001, and in response to the Rio Declaration, the European Commission prepared a proposal titled, *A Sustainable Europe for a Better World: A European Union Strategy for Sustainable Development.* Later, EU policy in sustainable development was outlined in the "6th European Community Action Programme" titled *Environment 2010: Our Future, Our Choice (2001).* In March 2005, the European Union relaunched the "Lisbon Strategy" and delivered a commitment to pursue and respect sustainable development as a "*key principle governing all the Union's policies and activities*".

This commitment was reflected in the approval by the European Council of the *Declaration on the Guiding Principles for Sustainable Development* (Council of the European Community, 2005). The Declaration incorporates *Policy Guiding Principles* such as the use of the best available knowledge, the precautionary principle and making the polluter pay. In addition, the *Key Objectives* and *Policy Guiding Principles* of the Declaration set out targets, indicators and an effective monitoring procedure in the Declaration to act as a basis for the sustainable development strategy. There is considerable overlap between the EU Action Programme and Declaration and the UK policy *Securing the Future*, in terms of giving effect to the Rio Declaration and Agenda 21.

The EU policy and the Declaration do not have binding force as a source of EU law in the United Kingdom. However, the policy and the Declaration represent a source of knowledge power for resolving public interest conflicts over sustainable development, where their application would facilitate the integrity of the administrative decision-making process in accordance with the relevant statute.

(c) The United States

The position in the United States is somewhat of a paradox. The United States has long been the global innovator of environmental legislation that has been the source and model for new legislation introduced in other countries around the world. For example, legislation dealing with environmental impact assessment, endangered species and contaminated lands. However, in marked contrast to the contemporary direction in legislative approaches to sustainable development in Australia and the United Kingdom, there is no equivalent Federal statute that incorporates sustainable development within its provisions. Furthermore, as Dernbach (2002b) states:

> "The United States has no national strategy for implementing sustainable development and no generally accepted indicators to mark progress along the way. Neither the executive branch nor the U.S. congress systematically analyse proposed activities to find ways to make significant progress on economic, environmental, social and security goals at the same time."

Dernbach (2002b) further identifies the failure of the US Congress and the Executive to systematically analyse proposed activities in terms of interrelating environmental, economic, social and security considerations. In this regard, Dernbach argues that a decisive step towards sustainability would be achieved in the United States if Government at all levels, Federal agencies and businesses adopted the leading sustainability practices of their national and international counterparts.

SUSTAINABLE DEVELOPMENT: LEGAL RIGHTS AS A SOURCE OF POWER FOR NEGOTIATING SOLUTIONS

Where sustainable development has been implemented into environmental legislation and policy, based on the Rio Declaration and Agenda 21, there would be common ground for its foundation to be dependent on ecological, economic and social and cultural considerations. The incorporation of sustainable development, in environmental legislation particularly, as well as an environmental policy, is a very significant source of knowledge power for parties seeking to rely on negotiation to resolve environmental conflicts. Such a case would arise where a civil enforcement action, e.g. for an abuse or misuse of statutory power in the administrative decision-making process, is contemplated as a trigger to facilitate the use of negotiation for conflict resolution.

The starting point would be to evaluate whether there may have been an error of law in the administrative decision made by a Government agency, for example whether the decision was authorised by the statute by taking into account the matters prescribed for the meaning of sustainable development as defined in the relevant legislation. Where any disparity can be found in the interpretation and application of the meaning for sustainable development, according to its meaning in the appropriate statute — or, in the appropriate case, the meaning in a policy, and the administrative decision — the threat of a potential judicial review challenge becomes real and not speculative. In this situation, it may become a basis for the agency to consider sharing its power to resolve an environmental conflict by negotiation where a BATNA analysis is undertaken as part of a conflict assessment procedure.

(a) Australia

ESD has been incorporated as a prescribed statutory object or provision in a number of Commonwealth environmental statutes, as well as for environmental and planning legislation in a number of States, and so represents a significant source of knowledge power in reviewing the administrative decision-making process by a Government agency. The legislative meaning for ESD sets out the matters that must be considered as a legal obligation, in the administrative decision-making process by a Government agency, for the granting of an authority, licence or permit under environmental legislation for a proposed action. A review of the factual and legal elements of the agency decision, in relation to the grounds for a judicial review, for example, would indicate whether a judicial review challenge could be sustained.

Where a particular statute, that is central to the outcome of an environmental or planning conflict, prescribes ESD (or any of its many variants in Australia, such as ecologically sustainable use) as a statutory object or provision, but does not provide any definition of its meaning, then regard could be given to a policy for its meaning for ESD, e.g. the *National Strategy for Ecologically Sustainable Development (1992).* However, for a policy to be regarded as a relevant consideration with respect to the meaning of ESD, it must be seen as facilitating the integrity of the administrative decision-making process and to be consistent with the purpose of the particular statute: see *Re Drake and Minister for Immigration and Ethnic Affairs (No. 2)* (1979) 2 ALD 634.

In addition, where a matter, such as ESD, is not provided in a statute it may be a relevant consideration for administrative decision-making by Government agencies if the subject matter, scope and purpose of the statute show the matter not to be an extraneous consideration: *Minister for Aboriginal Affairs v Peko Wallsend* (1985) 162 CLR 24. The reasoning of the High Court in the *Peko Wallsend case* was later relied upon by Stein J in *Leatch v Director-General of National Parks and Wildlife* (1993) 81 LGERA 143 to give effect to the precautionary principle that was contained in a national policy — but, at that time, the precautionary principle had not been incorporated into the relevant statute — in this case, the *National Parks and Wildlife Act 1974 (NSW)*.

Finally, a number of statutes introduced for the regulatory control of environmental harm introduced by some of the States and Territories, e.g. s. 5, *Environmental Protection Act 1994 (Qld)* and s. 10(2) *Environment*

Protection Act 1993 (SA), impose a public duty that relates to achieving the object of the Act — an object that includes achieving ESD.

(b) The United Kingdom

The *Planning and Compulsory Purchase Act (2004)* is a significant source of power for parties seeking to rely on negotiation to resolve sustainable development conflicts. The *Planning and Compulsory Purchase Act* makes it mandatory for regard to be given to national policies, e.g. "*Securing the Future. Delivering UK Sustainable Development Strategy*" (2005). Consequently, the five guiding principles in this policy represent relevant considerations for the administrative decision-maker to take into account in the decision-making process in order to achieve sustainable development under this statute.

The *Environment Act (1995)* and the *Local Government Act (2000)* also have statutory objectives directed towards sustainable development and so represent a significant source of power for parties seeking to resolve environmental conflicts under both these statutes, by negotiation. However, neither statute defines the meaning of sustainable development. In turn, this raises the status of the meaning of sustainable development in the national policies for administrative decision-making. There is a long-held principle of English public law that Ministers will follow their policies unless there is a reason not to do so: a policy is a foundation for fairness as it can facilitate the integrity and consistency in administrative decision-making (Lord Justice Sedley, 2004). Accordingly, an arguable case could be raised that the national policy, *Securing the Future. Delivering UK Sustainable Development Strategy* may be a relevant consideration for administrative decision-making by Government agencies as it sets out the matters that must be considered in decisions involving sustainable development. Furthermore, it follows that in order to be consistent with established English public law principles, the national policy would have to be seen to facilitate the integrity and consistency of administrative decision-making for sustainable development by the Government agency.

(c) The United States

There is no effective legal framework central to environmental regulatory control for sustainable development at the Federal legislative level. In the absence of any Federal environmental legislation that incorporates statutory provisions for sustainable development, the question of legal rights that may be pursued at the Federal level to challenge the legality of administrative decisions made, or actions taken, by a Government agency over sustainable development does not arise. At this stage, it would seem that in the United

States, at least, environmental conflicts over sustainable development will most likely arise in the States that have incorporated provisions for environmental regulatory control of sustainable development in legislation; alternatively, where sustainable development has been implemented in environmental policy.

ACHIEVING SUSTAINABLE DEVELOPMENT: SCIENTIFIC METHODOLOGY

A methodology for achieving sustainable development consistent with its meaning and application as defined in international declarations, environmental legislation and policies, must necessarily be based on multiple objectives: ecological, economic and social (including cultural) objectives.

(a) The Sustainable Development Balance

Achieving sustainable development requires a balancing of multiple objectives. In some conflict situations the question of finding the balance for sustainable development may be relatively straightforward as the balance may be clearly offset towards ecological considerations, e.g. sites having a high conservation status such as a World Heritage site or a site of spiritual significance for Indigenous people. In other conflict situations, the balance may be clearly offset towards socio-economic considerations, e.g. where the site is no longer pristine.

The balancing of ecological, economic and social and cultural objectives is dependent on there being a number of alternatives (or scenarios) that would all need to be evaluated. The most probable outcome for resolving a sustainable development conflict by negotiation would be a compromise solution as the parties would need to make concessions in relation to specific factual issues in dispute. In these circumstances, finding the balance for a sustainable development conflict would be facilitated by having a number of alternatives to evaluate along a continuum of sustainable development, rather than having the evaluation based on only one alternative. Reaching consensual agreement would be problematic if there were only one sustainable development alternative to evaluate.

(b) Multi-objective Analysis (or "Multiple Criteria Decision Aid Methodology")

Where multiple objectives are to be evaluated, there is also a complementary need for the evaluation to be based on objective criteria — a key element of principled negotiation (Fisher *et al.*, 1991). In these circumstances, *Multi-*

objective Analysis (or, as it is often referred to, *Multiple Criteria Decision Aid Methodology*) can be used as a decision-making aid for sustainable development conflicts (Figure 5.1).

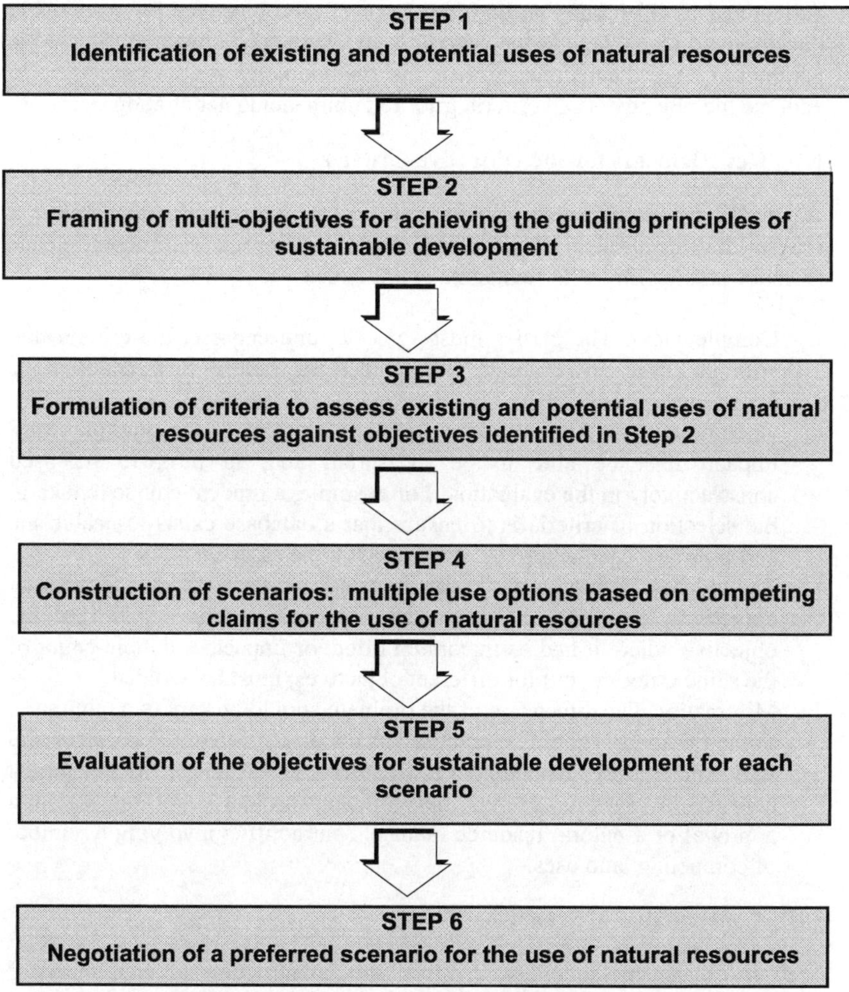

Figure 5.1 Sequential steps for evaluating the sustainable use of natural resources using multi-objective analysis (or "multiple criteria decision aid methodology") (based on Christie, 1993b)

The "drivers" for the methodology relate to the framing of the multiple objectives (Step 2) and the criteria that are to be used to evaluate each objective (Step 3). The criteria are objective and are chosen to provide some quantitative measure of the relevant objectives for sustainable development. Agreement by all parties on both the form of the objectives for sustainable development (Step 2), and the criteria to evaluate each objective (Step 3) is paramount. A critical dimension of the methodology is for the parties to endorse the objectives and criteria prior to commencing negotiations.

(c) Key Elements for the Objective Criteria

A number of authors have summarized the key elements for the objective criteria that are used to evaluate environmental impacts in time and space (Keeney and Raiffe 1976; Lahdelma *et al.*, 2000):

i. Completeness: The parties must agree to, and endorse, the criteria that will be used to evaluate the objectives before any negotiations commence.
ii. Operationality: Criteria selected as the end-point of the environmental impact must be able to be measured, and, in turn, to be used constructively in the evaluation. For example, a prudent course to take in the selection of criteria is to ensure that a database exists to enable the end-point for each environmental impact to be assessed.
iii. Redundancy: Where a particular criterion overlaps across a number of objectives, it should be evaluated for one objective only, i.e. the objective where it had its maximum effect or impact. A double-count of the same criterion, but for different objectives, must be avoided.
iv. Minimality: The dimension of the problem should be kept to a minimum. Some flexibility should be practised as the dimension of the problem will vary enormously depending on whether the sustainable development conflict involves a discrete development project requiring planning approval or a natural resource management conflict involving a number of competing land uses.

(d) Construction of Scenarios

Step 4 of the methodology involves the construction of a number of scenarios along a sustainable development continuum. A scenario is a hypothetical construction of the conflict, e.g. a proposed development project or a natural resource management land use problem, in which different weight or balance is given to ecological, economic and social objectives in each scenario. It is clear that an innumerable number of scenarios could be constructed for any sustainable development conflict. As Lahdelma *et al.*

(2000) point out, in many situations the number of alternative scenarios may be infinite, but the dimension of the conflict that will be the subject of the decision-making process requires a finite number of scenarios to be constructed. It also needs to be recognised that the continuum of scenarios is not fixed as the number may change during the process of evaluation.

(e) Evaluating Scenarios: Information Needs

Step 5 of the methodology involves the evaluation of the scenarios. It is generally recognised today that it would be rare to have a perfect information base for all the ecological, economic and social and cultural considerations central to any conflict over sustainable development. There will always be scientific uncertainty in the information base, especially for some long-term impacts on the physical and biological resources, including natural and cultural heritage values. Where there is scientific uncertainty, the traditional approach is to seek independent expert opinion on the issue. Such a step requires conclusions to be reached which are consistent with generally accepted scientific theories and research methods. Expert opinion, in these circumstances, is an important complement to the existing information base.

Consequently, multi-objective analysis is not intended to establish scientific truth where uncertainty in relation to environmental impacts exists. Rather, its purpose is to be an aid to environmental decision-making. Where appropriate, it may also impose specific conditions based on sound scientific and planning principles to offset any uncertainty arising from an imperfect scientific and technical database (Christie, 1993b).

(f) Selection of the Preferred Scenario

Step 6 of the methodology involves the selection of a preferred scenario. Lahdelma *et al.* (2000) describe the following alternative pathways that may be considered in selecting a preferred scenario:

i. Selection of one (or more) scenarios as the scenario that represents the best alternative. However, such an approach would diverge from the aim of the methodology to be an aid to decision-making if it was used by the administrative decision-maker to determine the decision end-point;
ii. Ranking the scenarios in the order that each scenario is considered to represent the best alternative. This approach provides more flexibility as it enables all scenarios to be considered and for the parties to agree to a scenario, independent of its ranking, giving reasons for their choice; and
iii. Acceptability analysis of the scenarios by describing what kind of preferences would give the best rank, or any specific rank, for each

alternative. This approach provides the parties with the maximum freedom for choosing a preferred scenario.

INCORPORATING MULTI-OBJECTIVE ANALYSIS (MULTIPLE CRITERIA DECISION AID METHODOLOGY) INTO THE ENVIRONMENTAL CONFLICT RESOLUTION PROCESS

Multi-objective analysis (Figure 5.1) was used to evaluate the competing claims for ESD and existing and future uses of natural resources in a public inquiry in Australia (Commission of Inquiry into Shoalwater Bay, Capricornia Coast, Queensland, 1994b). The background and scope of this public inquiry have been discussed in Chapter 2. There were four multiple objectives based on (i) ecological, (ii) economic, (iii) social objectives and, given the past Indigenous association with the land, (iv) an Indigenous interests objective.

The key features of the multi-objective analysis methodology used in the public inquiry process were as follows:

(a) Procedural Fairness and Trust in the Public Participation Process

It was important that the parties saw the multi-objective analysis process as one which was procedurally fair for everyone and was carried out in a transparent manner. The following aspects were adhered to from the outset:

i. All parties had equal access to all of the sources of information that would be used by the Commission in its decision-making role to make recommendations to Government on the future use of the area;
ii. A form of data mediation was also incorporated into the procedures to ensure all relevant information had been identified and made available to all parties before public hearings commenced; and
iii. The multi-objective analysis methodology and the objective criteria proposed to be used to assess ESD for competing land use claims were prepared by the Commission as a *Draft Public Discussion Paper* and written submissions sought from the parties. These submissions were then reviewed by the Commission and incorporated into the *Final Public Discussion Paper* and became part of the public inquiry procedures. The *Discussion Paper* emphasised that the methodology was to assist in the decision-making process and was not the decision end-point.

(b) The Multi-Objective Analysis Methodology: Evaluation of Multiple Competing Land Use Claims

It is of particular significance that the same objectives and criteria were used to evaluate all ten existing and proposed uses for natural resources and ESD, rather than using different criteria for a specific natural resource use to evaluate each objective. For example, the use of the study area for military training and national security purposes was assessed against the same objectives and criteria as those used to assess conservation, sand-mining and commercial fishing. This procedure represented a systematic and consistent approach to evaluating each objective for ESD.

(c) Selection of Objective Criteria to Evaluate the Objectives

The final set of criteria used to assess the compatibility of the competing uses of natural resources in the study area, with the multiple objectives for ecologically sustainable development, are summarised in Table 5.1. Criteria were chosen on the basis of operationality, non-redundancy and minimality. The aim was to avoid problems associated with using unnecessarily complex criteria, or criteria which could not be evaluated because of the absence of a suitable information base. Moreover, the evaluation was based on each objective, and all of the criteria, having equal weight.

(d) Construction of Scenarios

The scenarios used were a hypothetical construction of different combinations of all existing and proposed uses of natural resources. The scenarios were constructed in such a way that all competing uses of natural resources were incorporated in at least one scenario. It must be emphasised that it is the testing and evaluation of a specific natural resource use in each scenario that are of primary importance, rather than the actual structure of the scenario.

In terms of the issue of minimality and the dimensions of the problems of time and space, the spatial area of land and sea was the actual area of 502,000 ha used for military training purposes. In addition, the dimension of time was based on land uses which were known to be finite, in this case, the projected duration of 20 to 25 years for a sand-mining operation.

Four scenarios were constructed along a continuum of sustainable development. Each alternative scenario represented a different balance between ecological, economic and social considerations; Indigenous interests ran across all scenarios. Each alternative scenario incorporated all competing uses of natural resources in which the intensity of a human use varied over a very wide spectrum of land use (Box 5.1).

Table 5.1 Objective criteria used to assess the compatibility of competing uses of natural resources with ecologically sustainable development (Christie, 1993b)

Objectives		Criteria to Evaluate the Objective
Ecological *(Resource Management)*	C_1	Biological diversity
	C_2	Water removal/acquifer recharge balance
	C_3	Ecosystem resilience
	C_4	Water quality
	C_5	Feral or exotic species
Ecological *(Heritage Preservation)*	C_6	Protection/maintenance of World Heritage Convention values for the Great Barrier Reef Area
	C_7	Protection/maintenance of statutory criteria for placing the site on the Federal Register of the National Estate
Economic *(National and State)*	C_8	Gross domestic product
	C_9	Gross State product
	C_{10}	Household income (National and State level)
	C_{11}	Balance of payments of nation
	C_{12}	Employment/unemployment (National and State level)
	C_{13}	Non-market values, e.g. recreation, non-use values
Economic *(Regional Economy)*	C_{14}	Gross product
	C_{15}	Household income
	C_{16}	Employment/unemployment levels
	C_{17}	Range of employment options
	C_{18}	Range of infrastructure options
	C_{19}	Non-market values, e.g. recreation, non-use values
Economic *(Fiscal Consequences)*	C_{20}	Net fiscal consequences for the Federal Government
	C_{21}	Net fiscal consequences for the State and Local Government
Social *(National)*	C_{22}	Development and maintenance of national security
Social *(Social Equity)*	C_{23}	Community involvement in administrative decisions that affect them
Social Well-Being *(Regional)*	C_{24}	Quality of basic and special community services for residents
	C_{25}	The quality and provision of recreational and tourist facilities for residents and visitors
Indigenous Peoples Interests	C_{26}	Preservation and maintenance of Indigenous culture and heritage
	C_{27}	Promotion of Indigenous traditional knowledge for natural resource management
	C_{28}	Recognition of Indigenous interests in the distribution of benefits arising from the use of natural resources

(e) Evaluation of Scenarios

The criteria in Table 5.1 were used by the Commission to assess each objective, for each scenario, in order to identify the compatibility between each competing use of natural resources and the specific objective. The compatibility was characterised as "compatible", "non-compatible" or "uncertain"; reasons were given for these conclusions. No numeric scores (e.g. a score out of 10) were assigned to any criteria.

Written submissions were sought in response to the evaluation of scenarios published in the Interim Report. During the process of evaluation a new scenario was also constructed as the *preferred scenario* based on the best features of the original set of scenarios evaluated. The preferred scenario was then subject to evaluation using the objectives and criteria in Table 5.1.

(f) Opportunities for the Use of Multi-objective Analysis for Resolving Environmental Conflicts

Whilst multi-objective methodology is a well-accepted procedure used as a decision-making aid for public interest environmental and planning issues, it is most commonly applied to conflicts involving a single land use, such as a dam project, or a development activity requiring planning approval, e.g. a waste treatment plant. However, environmental decision-making involving the evaluation of sustainable development and multiple uses of natural resources is often a much more complex issue. The Shoalwater Bay public inquiry was an example of a complex public interest issue as it involved the compatibility with ESD of many competing land uses for natural resources. Whilst it could be demonstrated that multi-objective analysis methodology could be used as an aid for environmental decision-making on ESD, this was subject to one qualification: the need for and adequate data and information base to evaluate the multiple objectives for ESD was crucial.

The conclusions of Dyer *et al.* (1992) are particularly relevant as to the wider opportunities of the methodology for resolving environmental conflicts. The authors state that because most public sector problems involve multiple conflicting objectives — whether in environmental policy, water resources, energy or public health — the opportunity for the methodology is unlimited. In this regard, it could also be concluded that the methodology could be applied to the collaborative fact-finding and problem-solving procedures of ADR and negotiation to facilitate finding solutions for sustainable development conflicts.

Box 5.1 Summary of the general features of scenarios constructed along a continuum of sustainable development (Commission of Inquiry into Shoalwater Bay, Capricornia Coast, Queensland, 1994b)

Scenario 1: The Rights of One Resource User to Prevail over All Others
A multiple use scenario constructed around one particular resource use activity being permitted at a much greater level of use compared with the other. Sand-mining was maximised with major reductions in military/defence training operations.

Scenario 2: Steady Growth in Levels of Use over Time
A multiple use scenario constructed around some reduction in military/defence training operations to accommodate some sand-mining and some further mining exploration.

Scenario 3: Status Quo
A multiple use scenario in which only the existing natural resource uses were permitted — and only at the existing levels of resource use — but with some provision for future growth. This was a "benchmark scenario" as the central question became whether the status quo was the ideal mix for use of natural resources over the finite time period of 20–25 years.

Scenario 4: Preservation
Preservation prevailed as human use activities were limited. The use of the entire area was restricted to use as a conservation and wilderness area.

SUSTAINABLE DEVELOPMENT AND INDIGENOUS PEOPLE: A SPECIAL CHALLENGE FOR CONFLICT RESOLUTION

The Rio Declaration recognised that, in order for Indigenous people and their communities to achieve sustainable development, knowledge of their traditional practices was required together with a process in which they could participate effectively (Principle 22). In recognizing that "*respect for indigenous knowledge, cultures and traditional practices contributes to sustainable and equitable development and proper management of the environment*", the United Nation's Declaration on the Rights of Indigenous Peoples (2007) stated that:

> "Indigenous people have the right to participate in decision-making matters that would affect their rights, through representatives chosen by themselves in

accordance with their own procedures, as well as to maintain and develop their own indigenous decision-making institution (Article 18)".

The traditional approach to legislatures in many countries has been to introduce statutes that deal with individual rights of Indigenous people, e.g. civil rights, that extend to legal rights relating to land claims, protection of sacred sites, water rights and hunting and fishing rights. In addition, national park legislation provides legal rights relating to the preservation of the natural and cultural heritage of Indigenous people. Notwithstanding the enormous significance of such statutes for Indigenous people, they do not necessarily prescribe sustainable development as a matter that must be taken into account by a court as a legal obligation.

The central issue is how best to effectively apply Indigenous traditional knowledge in resolving conflicts over sustainability. Any failure to do so will lead to limitations in the available information for the administrative decision-making process. The potential for uncertainty in identifying potential adverse environmental impacts would increase. In the extreme case, litigation would be inevitable. The challenge is considered in terms of the conflict resolution process and legal rights that would provide Indigenous people with an adequate basis of power to participate effectively.

(a) Litigation or Negotiation?

Deloria (1996) points out that relations between Indigenous peoples and Government arrived at by negotiation can build mutual understanding and trust — even in volatile situations. Moreover, through good faith dialogue, it is possible to reach agreements that provide a sense of legitimacy for all concerned. Litigation, on the other hand, by prohibiting the participation of Indigenous peoples, is less attractive for customary behaviour, less coherent, less efficient and often contentious (Frickey, 1997).

(b) Approaches to Participatory Management

Goodman (2000) recognises that there are two sources of conflict arising over the legal rights of Indigenous peoples for the co-management of natural resources: information, given the uncertainties of science, as well as values. In terms of the requirements for a dispute resolution process for resolving conflicts over a co–management regime, Goodman (2000) considers that the process should be clear and objective to ensure a fully participatory management framework. In this regard, the shared responsibility for fact-finding and problem-solving procedures, which are the foundations of ADR, would provide advantages for ADR, relative to litigation, to manage and to resolve conflicts over sustainable development and co-management of natural resources. In addition, with appropriate intervention procedures, conflicts over values can be addressed in the ADR process. Furthermore, a participatory approach to co-management of natural resources that provided Indigenous peoples with a significant role in the preparation and ownership of solutions would have positive benefits for their implementation.

Goodman (2000) considers that the following approaches would facilitate effective participatory management:

i. *A court sanctioned decree with continuing judicial oversight:* The approach involves a detailed negotiated agreement being developed in advance, outside of litigation, that addressed specific shared management issues for the use of natural resources that relate to sustainable development. The agreement would be flexible so as to deal with conflicts arising from potential unforeseen situations on sustainability that may arise over time. The agreement would be prepared under a "rights-based approach" to conflict resolution and so could be enforceable as a consent decree under the ongoing jurisdiction of the appropriate court.
ii. *Consultation with Indigenous communities as an expert agency:* This approach requires the Government agency to recognise the Indigenous community as the expert agency in evaluating aspects of the use of natural resources that relate to their legal rights, e.g. hunting and fishing rights, the protection of cultural resources. In this regard, Indigenous people would be in a significant position to share power with a Government agency to pursue this approach where a statute contains a provision, such as the Australian statute, the *Environmental Protection and Biodiversity Conservation Act 1999 (Cth)*. One object of this Act is "*to promote the use of indigenous peoples' knowledge of biodiversity with the involvement of, and in co-operation with, the owners of the knowledge*": Section 3(1)(g).

Formal consultation by the Government agency with the Indigenous community, under this approach, would result in an outcome in which the Indigenous community would provide expert opinion based on traditional knowledge. This opinion would be prepared in the context of whether a proposed development would be consistent with sustainable development, would be consistent with the proper management of the environment and would not adversely affect Indigenous legal rights to use natural resources.

(c) Statutory Rights: Sustainable Development and Co-management of Natural Resources

Environmental legislation for the use of natural resources that prescribes both sustainable development and co-management opportunities for Indigenous peoples, as matters that must be taken into account in the administrative decision-making process by a Government agency as a legal obligation, represents a significant source of power for Indigenous peoples for conflict resolution. In this situation, their legal rights would extend to the much wider rights related to ecological, economic and social (including cultural) sustainability. Positive social and economic benefits would be additional benefits for natural and cultural heritage values because of the sustainable development concept. One such example of these statutory rights is the Federal statute in Australia, the *Great Barrier Reef Marine Park Act 1975 (Cth).* Sustainability is one statutory object for management plans, as the aim is "*To ensure that activities within areas of the Marine Park are managed on the basis of ecologically sustainable use*": Section 39Y(c).

In addition, the statute contains the following provisions for joint management of the Marine Park (at s. 39ZA):

> "(1) The [Great Barrier Reef Marine Park] Authority may enter into an agreement or arrangement ... with a group of People who are representative of **a community group that has a special interest** in an area of the Marine Park.
>
> (2) The agreement or arrangement may relate to the development and/or the implementation of a plan of management ... within the area concerned and may if the Authority considers appropriate, provide that, if such a plan of management is prepared, the community group is to manage the area ... jointly with the Authority in accordance with the plan (Author's emphasis)".

The meaning of the phrase "a *community group having a special interest*" is defined in Section 39V:

> "A reference ... to a **community group having a special interest** in an area of the Marine Park includes a reference to the people in the group who have some form of native title to the area or its resources or have some other special identification with the area or its resources" (Author's emphasis).

Statutory rights, such as these provisions in the *Great Barrier Reef Marine Park Act*, represent a pathway for Indigenous people to pursue opportunities for co-management of natural resources and sustainable development through negotiation. Statutory rights are a source of knowledge power enabling the challenge for sustainable development, as defined in the Rio

Declaration and Agenda 21, to be taken up using the participatory approaches advocated by Goodman (2000) and interest-based negotiation.

FINDING SOLUTIONS FOR SUSTAINABLE DEVELOPMENT CONFLICTS: LITIGATION v NEGOTIATION

The approach of courts and litigation to find solutions for sustainable development conflicts can be distinguished from that for ADR and negotiation. For example, under planning legislation, a court has to consider whether the proposed use for the land that is the subject of the application is consistent with the legal meaning of sustainable development in terms of its meaning as defined in the relevant statute. Sustainable development would be considered in the context of the land use being proposed, but only for a single or discrete land use. The issue for the court to determine may not necessarily be whether the proposal is the best possible use, but whether it is a suitable or acceptable use in environmental terms. The only flexibility for the court would be to determine whether approval should be given, with or without conditions, to ensure compliance with sustainable development.

In contrast to courts and litigation, ADR enables an evaluation of a number of scenarios along a sustainable development continuum in which the balance between ecological, economic, social and cultural considerations varies, rather than attempting to provide a single definitive solution based on the evaluation of one scenario to resolve the conflict. Consequently, the use of ADR to find a solution for a sustainable development conflict has a number of advantages that reflect considerable scope in flexibility during the negotiation process:

i. ADR enables the evaluation of a range of scenarios along a sustainable development continuum in which the balance between ecological, economic and social and cultural considerations varies;
ii. ADR has the potential to formulate new scenarios, as negotiations proceed, and which may be advantageous for finding consensus between the parties; and
iii. ADR allows the construction of a preferred scenario. However, it must be clearly recognised that any new or preferred scenario can only be formulated after negotiation and the ADR process commences; such an alternative scenario may not have been envisaged, from the outset, by the parties.

Having such flexibility potentially available for the parties through the ADR process, whereby the number of scenarios can change after negotiations commence — including the construction of new scenarios — is a significant advantage for ADR over the courts and litigation for finding solutions for sustainable development conflicts (Christie, 1993c). Moreover, such an advantage would be facilitated where multi-objective analysis and the creation and evaluation of scenarios were also a central element for joint problem-solving — a key procedural feature of ADR, but not the courts and litigation.

Finding solutions for sustainable development conflicts requires an evaluation of multiple objectives over a finite period of time. A further advantage for ADR, relative to courts and litigation, is that only ADR can provide solutions that are both prospective, i.e. designed to anticipate future policy or practical conflicts over sustainable development, and retrospective, i.e. reactive to the needs for sustainable development and the existing issues in dispute.

6. Environmental impact assessment

"The purpose of the EU directive on EIA was to ensure that planning decisions which may affect the environment are made on the basis of full information."

Lord Hoffmann in *R. v North Yorkshire County Council ex parte Brown* [2000] 1 A.C. 397

"The average accuracy of quantified, critical, testable predictions in environmental impact statements in Australia to date is 44 ± 5% ... At present, in Australia at least, our predictions are less than 50% accurate on average and over two orders of magnitude out on occasions."

Buckley (1991)

"Developers ... have quite defined priorities with respect to ... land ... They generally have fairly clear ideas about what it is they want to do, when they want to do it and why. Negotiations will revolve around how the [development] can be made to fit within existing patterns of land use and development ... [However] negotiations could operate to an agenda that emphasized, for example, cultural issues, community development and the preservation of ecologies. Proposed development could be discussed in terms of its consistency with Indigenous aspirations on these issues."

Mick Dodson (1996), Aboriginal and Torres Strait Island Social Justice Commissioner

INTRODUCTION

The United States statute, the *National Environmental Policy Act of 1969* ("NEPA"), signed into law on 1 January 1970, was an innovative and pioneering statute. Craig (2005) refers to NEPA as possibly being "*the most successful legal export in history*" as it has been a model for over 100 countries. NEPA required Federal agencies proposing to undertake actions that may have a "significant" effect on the environment, to prepare an environmental impact statement ("EIS"). The term "EIS" is also used in Australia. In the UK, the equivalent document is termed an environmental statement ("ES"). The US Supreme Court in *Baltimore Gas & Elec. Co. v Natural Resources Defense Council*, 462 U.S. 87, at 97 (1983), identified the twin aims of NEPA: an obligation on an agency to consider every significant aspect of a proposed action and to ensure that the agency informs the public that it has considered environmental concerns in the decision-making process.

An EIS is a carefully researched report which identifies the likely or possible environmental consequences of a proposed development or activity or policy, alerting the government, developer and the community, as fully as possible, to any environmental risks associated with the proposal. The process of reviewing the EIS is termed environmental impact assessment ("EIA"). The EIA process is the systematic appraisal of the positive and negative effects on the environment that are possible, or likely to arise, from a development proposal or activity. The EIA process is undertaken independently of the proponent of the development or activity, i.e. the party who may have submitted the EIS to the Government agency. The purpose of the EIA is to assist the administrative decision-making process of a Government agency. The administrative decision by the Government agency may take the form of an approval, deferment or rejection of the proposal, or approval of the proposal with conditions that incorporate measures to mitigate adverse impacts. These conditions then become part of the developmental consent or permit (Gilpin, 1986).

Concerns over the adequacy of the EIA process have long been expressed in the scientific literature. A 1976 editorial in *Science*, the prestigious journal of the American Association for the Advancement of Science, stated:

> "The idea (an EIS) designed to protect our natural resources ... has backfired ... Someone is inevitably available to receive these funds, conduct the studies regardless of how quickly results are demanded, write large, diffuse reports containing reams of uninterpreted and incomplete descriptive data, and in some cases, construct 'predictive' models irrespective of the quality of the database ... impact statements seldom receive the hard scrutiny that follow the publication of scientific findings in a reputable scientific journal" (Schindler, 1976).

Some 13 years later, Fairweather (1989) claimed that the situation for EIA in Australia was the same as, or worse than this United States analysis. In addition, Stewart and McColl (1994) regard the analysis of economic and social issues as continuing limitations in the effectiveness of EIA procedures. Some of the key concerns of science for EIA as an aid for decision-making have been identified (Buckley, 1989; Resource Assessment Commission, 1993). Where present, these concerns represent significant shortcomings for good environmental outcomes:

- The integrity of the EIS, as there may be the possibility for biased and improper statements being prepared;

- Limitations imposed by the accuracy of impact predictions. This problem is exacerbated if there is no public access to environmental monitoring data and so no means for the public to assess whether the predicted environmental impacts were correct;
- A lack of integration with strategic management;
- Potential cumulative impacts over time from multiple projects, as well as long-term impacts, may not always be considered;
- Perceptions of inordinate weighting of financial and economic benefits, relative to potential unfavourable impacts on communities and the environment;
- Following project approval, Government agencies responsible for regulatory control of project conditions to monitor impacts or to comply with mitigation measures may not be directly accountable to the public. There may be no mechanism to ensure Government agencies are fulfilling their obligations; and
- Inadequate public participation procedures. Opportunities for the public in actual decision-making in the EIA process are limited.

Concerns expressed over the adequacy of EIA are not restricted to science as concerns from a legal perspective, such as excessive litigation, have been raised. Over the first 12 years of the NEPA in the United States, around 15,000 EISs were prepared. During this time, 1445 NEPA law suits were filed for Federal proposals (Gilpin, 1986). From a legal perspective, the complexity and sheer volume of prescribed procedures in the NEPA and Regulations for undertaking an EIS are reflected in the following observation by France (1990). It provides a sobering reminder of how prescribed statutory procedures for undertaking an EIS in these circumstances may diminish the effectiveness of the EIS as a decision-making aid:

> "After twenty years of experience with NEPA [in the United States], a process has evolved ... that does not work ... Continuous litigation has caused NEPA to evolve into a highly legalistic set of agency procedures. ... NEPA has evolved from a decision-making aid to a bureaucratic process that only incidentally aids agency decision-making. The courts have forced federal administrators to produce environmental studies that the administrators think are useless."

Concerns of Indigenous people are also particularly relevant as land that is in the resources sector and under traditional ownership is frequently the subject of developmental proposals. In Australia, one Government agency (Department of Families, Youth and Community Care, 1996) has made the following analysis of Indigenous people's concerns over the EIA process:

- Lack of involvement of Indigenous people and their representative organisations in the EIA process;
- Non-Indigenous determination of cultural heritage significance and social impact;
- Lack of direct participation by Indigenous people in cultural heritage assessment;
- Poor identification of social impacts, including socio-economic concerns; and
- Indigenous people's knowledge, concerns and potential contribution to environmental management and quality frequently being overlooked.

Notwithstanding all of the above concerns, one inevitable fact has emerged over time: environmental assessment of developmental proposals, in one form or another, is here to stay. The challenge for EIA, as a process to assist administrative decision-making, is to reconcile the concerns of science for environmental protection and management with the needs of law for adherence to procedural obligations imposed by environmental assessment legislation for undertaking an EIS. Superimposed on this challenge is the need to effectively involve the community and Indigenous people in the EIA process in a procedurally fair manner. The public participatory approach needs to involve the community and lead to an outcome where anyone having a recognised interest in the proposed action is able to emerge from the EIA process with some sense of gain and a sense of ownership in the administrative decision.

EIA AND SUSTAINABLE DEVELOPMENT

The EIA process has long-term implications for sustainable development for any proposed development or activity "*because sustainable development depends on protecting the natural resources which is the foundation for further development*" (Gilpin, 1995). EIA is seen as a planning tool that integrates environmental considerations into decision-making at the project, programme and policy levels. Ultimately, the focus is on an integrated assessment of environmental, economic and social effects of new proposals to ensure better informed administrative decision-making in support of sustainable development (Abaza *et al.*, 2004).

THE SCOPE AND PURPOSE OF AN EIS

In contrast to environmental protection legislation where the focus is on the regulatory control of some component of the environment, such as air or biodiversity, environmental assessment legislation sets out procedural obligations that must all be complied with in the administrative decision-making process undertaken by a Government agency (Ferrey, 2004). The procedural obligations cover a wide spectrum of the administrative process ending with the final administrative decision made by the Government agency to approve or reject the proposal. All administrative decisions made on procedural points along the sequential stages of the EIA process have the potential to be the subject of a judicial review challenge, not only the final administrative decision made by the Government agency (Weston, 2002). Some of the key procedural obligations prescribed under legislation, at various stages of the EIA process, include:

i. A key procedural obligation and one that is the crux for the EIA process is the legislative trigger which require a proposed action or project to be subject to an EIS, or some other alternative form of environmental assessment. A common trigger for the administrative decision that must be made, in Australia, the UK and the United States, is the likelihood for the proposal to have a "significant" impact on the environment. Consequently, the legal meaning and the factual basis for the term "significant" are crucial for any administrative decision made for each particular proposal.
ii. Other procedural obligations imposed by legislation relate to the substantive content and issues that will be the subject of the EIS, questions of statutory interpretation, e.g. the meaning of "significant", the form of the environmental assessment, the scope and nature of public participation in the EIA review process and procedures in the administrative decision-making process to approve or reject the proposal.

(a) The EIS and Decision-Making

An EIS is an aid to administrative decision-making by a Government agency. In the United States an EIS is to serve practically as an important contribution to the decision-making process and cannot be used to rationalise or to justify decisions already made: C.F.R. Section 1502.5. The true purpose of the EIS is not to make an exhaustive document in and of itself, but to inform decision-makers about the environmental consequences of their action: *Salmon River Concerned Citizens v Robertson* 798F. Supp. 1434 (E.D. Cal. 1992).

In the Australian case, *Prineas v Forestry Commission of New South Wales* (1983) 49LGRA 402, the following observations made by Cripps J support the legal approach in the United States:

> " … provided an environmental impact statement is comprehensive in its treatment of the subject matter, objective in its approach, and meets the requirement that it alerts the decision-maker, members of the public and the [Government agency] to the effect of the activity on the environment and the consequences to the community inherent in the carrying out, or not carrying out of the activity, it meets the standards imposed by the [EIA legislation] …
>
> An environmental impact statement is not a decision-making end in itself ... its purpose is to assist the decision-maker."

A similar situation exists in the UK. In *Berkeley v Secretary of State for the Environment, Transport and the Regions* [2001] Env. L.R. 16, Lord Hoffmann, in his decision, referred to the UK Government publication *Environmental Assessment: A Guide to the Procedures* (HMSO, 1989) and endorsed the following statement on the purpose of the EIA process:

> "At the same time it can help to inform the public on the substantive issues which the local planning authority will have to consider in reaching a decision. It is a requirement of the Regulations that the environmental statement must include a description of the project and its likely effects … "

Later, in the UK case, *Belize Alliance of Conservation Non-Governmental Organisations (BACONGO) v The Department of the Environment* [2004] Env. L.R., 38, the Privy Council majority (3:2) adopted the observations made by Cripps J in the Australian case, *Prineas v Forestry Commission of New South Wales.*

(b) Scientific Information and the EIA Process

Scientific information is a cornerstone for the EIA process and administrative decision-making. However, information deficiencies such as inaccurate, incomplete and inaccessible information make full access to information in the EIA process problematic (Sinclair and Doelle, 2003).

There is a legal distinction in the EIA process between resolving scientific conflict and statutory procedural obligations — such as informing the decision-maker and members of the public as to the environmental impacts that need to be recognised. In the United States case, *Jicarilla Apache Tribe of Indians v Morton* 471 F.2d 1275, 1280 (9th Cir. 1973), the court concluded that the purpose of the EIA process was not to reconcile divergent scientific opinion on a specific issue or to require that information be complete. The court observed that science cannot always provide final or definitive answers and to impose such a requirement could mean that the project proposed could never commence. Similarly, in the Australian case, *Warren v Electricity Commission of New South Wales* (1990) 130 LGERA 565, Cripps J stated:

> " ... it must be steadily borne in mind that the issue before me is not whether I am satisfied upon the present state of scientific knowledge there is a cause and effect relationship between ELF electric and magnetic fields and damage to human health ... What I am required to determine in these proceedings is whether the challenge has been successfully made out that the decision making process was [procedurally] flawed".

UK authorities have focused on the need for full information to be available in the EIA process for decision-making: *R v North Yorkshire County Council ex parte Brown* [2000] 1 A.C. 393. The question of whether an environmental statement prepared for a waste treatment and recycling facility had fully provided all relevant information was considered in *Hereford Waste Watchers v Hereford Council* [2005] EHWC 191 Admin. The High Court concluded that the planning permission was flawed as: "*The information which the planning authority requires and which it has stipulated should be made available prior to the development commencing, should have been made available prior to the planning permission being granted.*"

In *Berkeley v Secretary of State for the Environment, Transport and the Regions* [2001] Env. L.R. 16, Lord Hoffmann referred to his earlier comments in *R. v North Yorkshire County Council*, and emphasised the "*directly enforceable right of the citizen*" to participate in the EIA process and approved the following statement in the UK Government publication "*Environmental Assessment: A Guide to the Procedures*": " ... *One of the aims of a good environmental statement should be to enable readers to understand for themselves how its conclusions have been reached, and to form their own judgments on the significance of the environmental issues raised by the project.*"

A *Commentary* on this case, reported at the end of the decision by the House of Lords ([2001] Env. L.R. 16), notes that an important part of the EIA process is to facilitate public comment by providing the public with all relevant information in a form that can be understood.

LEGISLATIVE APPROACHES AND THE EIA PROCESS

(a) Australia

The initial Federal statute for the EIA process was the *Environment Protection (Impact of Proposals) Act 1974 (Cth).* This Act, along with other Federal environmental legislation, became consolidated as part of the Federal statute, the *Environment Protection and Biodiversity Conservation Act 1999 (Cth)* (the "EPBC" Act) that came into force on 16 July 2000. EIA under this statute must be undertaken within the context of sustainable development as one of its objectives is "*to promote ecologically sustainable development*". The States and Territories, through their environmental and planning legislation, also have separate statutory schemes for EIA.

Proposed actions requiring environmental assessment
The trigger for some form of an environmental assessment under the EPBC Act is when an "action" has been identified. The statutory definition for action includes "*a project; a development; an undertaking; an activity or series of activities; as well as any alteration of any of these things*" (at s. 523). However, only a "controlled action" is subject to the environmental assessment and approval process. The administrative decision that a proposal is a controlled action is based on information ("referral information") for the proposed action provided by the proponent, or a State or Territory, as the case may be. Generally, the question of referral of an action to the Federal agency will be made by the proponent.

Controlled actions Controlled actions arise in three broad groups, or classes.

- *Group 1: Actions relating to matters of national environmental significance:* In this group, a controlled action is an action that (i) "*has, will have or is likely to have*" (ii) a "*significant" impact* on (iii) a "*matter of national environmental significance*". All three elements must be satisfied for the action to be a controlled action. The EPBC Act identifies those matters which come within the statutory meaning of "matters of national environmental significance" in Chapter 2, Part 3, Division 1 of the Act:

  i. World Heritage properties: s. 12;
  ii. National heritage places: s. 15;
  iii. Ramsar wetlands of international importance: s. 16;
  iv. Nationally threatened species and ecological communities: s. 18;

v. Listed migratory species protected under international agreements: s.20;
vi. Nuclear action, e.g. establishing or significantly modifying a nuclear installation or large scale disposal facility for radioactive waste, mining or milling uranium ore, transport of nuclear fuel or radioactive waste: ss. 21, 22;
vii. A Commonwealth marine area: ss. 23, 24;
viii. An action prescribed by Regulation: s. 25.

- *Group 2: Actions relating to land owned or leased by the Commonwealth*: *s.* 26: A controlled action is an action that (i) "*has, will have or is likely to have*" (ii) a "*significant*" *impact* (iii) on "*the environment on Commonwealth land*". This provision covers land owned or leased by the Commonwealth, e.g. Commonwealth reserves and national parks, defence lands.

- *Group 3: Actions relating to the environment inside or outside the Australian jurisdiction: s. 28:* A controlled action is an action by the Commonwealth or its agencies that (i) "*has, will have or is likely to have*" (ii) a "*significant*" impact (iii) on "*the environment inside or outside the Australian jurisdiction*".

Controlled actions and statutory proof Two key elements of proof imposed by the EPBC Act for satisfying the requirements for a controlled action are common to all three groups. The questions of law are whether the action (i) "*has, will have or is likely to have*" a (ii) "*significant*" impact.

The requirement that an action that "*has*", "*will have*" or "*is likely to have*" a significant impact results in a wide spectrum for proof. At one end of the spectrum (i.e. the action "*has*") is one in which the civil standard of proof, the balance of probabilities, applies for causing actual adverse impacts. At the other end (i.e. the action is "*likely to have*") the requirement for proof for the potential to cause adverse impacts is relatively less onerous. In *Jarasius v Forestry Commission of New South Wales* (1988) 71 LGRA 79, the meaning of the word "*likely* [to significantly affect the environment]" — as prescribed in the *Environmental Planning and Assessment Act 1979 (NSW)* — was considered. The court concluded that "likely" meant a "real chance or possibility" — not "more probably than not". Consequently, the wide legal requirements for proof of causation arising under the EPBC Act represent a significant source for knowledge power for a party seeking to use the threat of a civil enforcement action as the basis for managing or resolving an EIA procedural conflict by negotiation.

Controlled actions and significant impact Matters prescribed by the *Environment Protection and Biodiversity Conservation Regulations 2000 (Cth)* (the "EPBC Regulations") must be taken into account in deciding the question of whether an action is likely to have a "significant" impact on the environment. At this stage no such Regulations have been made under the EPBC Act. However, there is a long-held body of case law from environmental litigation in Australia that the term "significant" impact means an impact that has "*an important or notable effect on the environment*": *Tasmanian Conservation Trust Inc. v Minister for Resources* (1995) 55 FCR 516. However, Branson J in *Booth v Bosworth* (2001) 114 FCR 39, by reference to the NEPA of the United States, concluded that the question of a "significant" effect on the environment required a consideration of two broad factors — "context and intensity" — and defined a "significant impact" under the EPBC Act as "*an important, notable* [impact] *or of consequence having regard to its context or intensity*".

The Federal Government has now released policy guidelines that set out criteria for assessing "significant" impacts under the EPBC Act for actions impacting on matters of national environmental significance or Commonwealth land as well as for actions by Commonwealth agencies: *EPBC Policy Statement, Significant Impact Guidelines: 1.1 Matters of National Environmental Significance, 1.2 Actions Impacting on Commonwealth Land and Actions by Commonwealth Agencies (May, 2006).*

The forms of environmental assessment

Under Section 87 of the EPBC Act, a wide range of approaches may be used, varying from the use of a checklist approach to ensure all environmental issues have been considered, through to an EIS. A decision on the form the assessment is to take is made by the Minister following the provision of information ("*the referral information*") on the action by the person proposing to take the action ("*the designated proponent*"). The Minister must choose one of the following approaches, and must consult, before making the decision:

- Assessment on referral information;
- Assessment on preliminary documentation;
- Assessment by an EIS;
- Assessment by a Public Environmental Report; or
- Assessment by public inquiry.

Regulation 5.03A of the EPBC Regulations sets out the criteria for evaluating the referral information that are to be considered by the Minister in deciding the form of environmental assessment. The matters that must be addressed by the Minister in an EIS are prescribed in Schedule 4 of the EPBC Regulations.

There is also an additional environmental assessment approach under s. 87 of the EPBC Act — an "*accredited assessment process*", whereby the controlled action is assessed by an accredited State or Federal Government process with the State or Federal Government agency preparing the assessment report. Under s. 45 of the EPBC Act, the Minister has statutory power to make bilateral agreements with the States and Territories. Any such bilateral agreement entered into must be consistent with the prescribed statutory objects for such an agreement, e.g. "*to protect the environment*" and "*to promote the conservation and ecologically sustainable use of the natural resources*": s. 44. In regard to statutory interpretation, it should be noted that the term "conservation" is not defined in the EPBC Act; nor is there a statutory definition for "ecologically sustainable use". These issues are not only problematic for the administrative decision-maker, but also may prove difficult to establish with scientific evidence that is certain, complete or accessible. These statutory limitations emphasise the need for the participatory approach to involve the public in a meaningful manner in the EIA process in order to offset the potential threat of litigation — or perceptions of manipulation.

Public participation and the EIA process

Relative to the UK and the United States, the statutory provisions for public participation in the EIA process under the EPBC Act are not as extensive in scope. Assessment by referral information, preliminary documentation, EIS and by a Public Environmental Report restricts public participation to public comment. In contrast, the public inquiry process extends public participation by providing for public inputs, apart from public comment, and may include public hearings. However, the practical reality is that public environmental inquiries are used infrequently in Australia, today.

Amendments to the EPBC Act in December 2006 have expanded the opportunities for public consultation and access to information. A summary of public comments is now to be included in the final EIS. In addition, the Minister may seek comment on the assessment report as well as the conditions proposed for a project (Department of the Environment and Heritage, 2007).

Approval of a controlled action

Regardless of the form of assessment, the decision in relation to approval of the proposal is made by the Minister for the Environment following a Recommendation Report prepared by the Government agency being provided to the Minister (Department of the Environment and Heritage, 2007).

(b) The United Kingdom

The EIA process is based on environmental information provided by the proponent of the proposed project and a report in support of the proposal, the Environmental Statement. The Environmental Statement identifies potential significant adverse environmental impacts that may arise from the project and also sets out management practices and mitigation measures that may be used to offset these impacts (Bell and McGillivray, 2006).

The sources of law setting out administrative procedures for the EIA process that must be followed by a Government agency are contained in EU Directives and UK domestic law. The relevant EIA Directive is Directive 85/337/EEC *on the assessment of the effects of certain public and private projects on the environment* (the "EIA Directive"). It was initially amended by Directive 97/11/EC. The EIA Directive was a significant but somewhat controversial source of EU legislation as there were strong objections from some Member States. It was the first source of EU legislation on broad-based land issues. It also dealt with administrative decision-making procedures. It took almost 16 years before it was enacted in 1985 (Wolf and White, 1997). Following the ratification by the UK in 2005 of the Aarhus Convention (1998) for public participation, the EIA Directive was further amended by the Public Participation Directive (2003/35/EC) to provide, amongst other things, amending the EIA Directive "with regard to public participation".

The amended EIA Directive (Directive 97/11/EC) has been incorporated into UK domestic law, as Regulations made under the *Town and Country Planning Act 1990*: the *Town and Country Planning (Environmental Impact Assessment) (England and Wales) Regulations 1999* (the "TCP Regulations").

Projects requiring EIA

The trigger for the EIA process is to decide whether the proposal comes within the prescribed legal meaning of a "project" and, if this is the case, whether "consent" is required. Although the TCP Regulations refer to "developments", rather than "projects", Bell and McGillivray (2006) consider that the meaning of "project" in the EIA Directive reasonably approximates the meaning of "development" in English planning law.

Project and consent: legal meaning

The EIA Directive provides the following definitions for "project" and "consent":

- "Project" means the execution of construction works or of other installations or schemes. Project is also defined to mean other interventions in the natural surroundings and landscape including those involving the extraction of mineral resources; and
- "Development consent" means the decision of the competent authority, or authorities, which entitles the developer (i.e. the applicant seeking authorisation for a private project or the public authority which initiates a project) to proceed with the project — where "developer" means the applicant seeking authorisation for a private project or the public authority which initiates a project.

Listed projects and EIA

The EIA Directive and the TCP Regulations provide a list of projects where an EIA is required. There is considerable overlap between the description of projects in the Annexes of the EIA Directive and developments in the Schedules of the TCP Regulations.

Generally, an EIA is compulsory for large-scale projects that are listed in Annex or Schedule 1. Examples of Annex I projects (or Schedule 1 developments) which always require an EIA include installations (i) for the production or enrichment of nuclear fuel or for the final disposal of irradiated nuclear fuel or radioactive waste or its storage, (ii) waste disposal installations for the incineration or chemical treatment or landfill of hazardous waste and (iii) waste water treatment plants with a capacity exceeding 150,000 population equivalent.

Projects or developments that are listed in Annex II of the EIA Directive (and Schedule 2 of the TCP Regulations) will only require an EIA subject to the following qualification: whether the proposal is likely to have a significant effect on the environment. In addition, where a large-scale project does not come within the description prescribed in Annex I of the EIA Directive (or Schedule 1 of the TCP Regulations), it too will require an EIA if the project is likely to have a significant effect on the environment. There are also developments defined in the Schedules of the TCP Regulations that extend beyond the projects defined in the EIA Directive. Bell and McGillivray (2006) state that whilst, in principle, EIA under the TCP Regulations would be required for projects that were not listed in the EIA Directive, "*there is an apparent reluctance in practice to extend the scope of an EIA beyond that required by EC law*".

Defined projects and EIA

There are also a number of "defined projects" recognised as requiring an EIA, but which do not come within the meaning of development in domestic land use planning legislation as defined in the *Town and Country Planning*

Act. Defined projects have their own consent procedures that are prescribed in separate regulations, specific for each project (Stookes, 2005; Bell and McGillivray, 2006) and include:

- Oil and gas pipelines;
- Nuclear power stations, other power stations and overhead power lines;
- Forestry projects;
- Marine fish farms;
- Projects under the Birds and Habitat Directive.

Listed projects: assessment of significant impacts

The determination of whether an EIA is required for a Schedule II project is based on thresholds or selection criteria set by the Member States: Article 4(2)(3), EIA Directive. The threshold question is whether the project would be likely to have a significant effect on the environment. Annex III of the EIA Directive (and Schedule 3 of the TCP Regulations) sets out the relevant criteria that are to be taken into account, on a case-by-case basis. The selection criteria are subdivided into three broad groups which all must be taken into account by the administrative decision-maker in determining whether an EIA is required. The selection criteria characterise the potential significant impacts of a project in relation to the characteristics of the development, e.g. size, cumulative impacts, waste produced and the location of the development, e.g. the environmental sensitivity. The focus of the EIA Directive is on the following criteria:

- The extent of the impact (geographical area and size of the affected population);
- The transfrontier nature of the impact;
- The magnitude and complexity of the impact;
- The probability of the impact;
- The duration, frequency and reversibility of the impact.

Public participation and environmental assessment

It is particularly significant that amendments to the EIA Directive in 2003 (by Directive 2003/35/EC *providing for public participation in respect of the drawing up of certain plans and programmes relating to the environment*) also implement the obligations arising under the Aarhus Convention. Consequently, the EIA Directive now provides for a much wider scope and a more expansive approach for public participation in the EIA process, relative to other jurisdictions such as Australia. However, the public still only have a limited, direct influence on the administrative decision made by the Government agency under this Directive. Article 2(d) requires the administrative decision-maker to examine the comments and opinions expressed by the public and to then inform them about the decision and the reasons and considerations upon which it was based.

(c) The United States

Federal agencies are required to prepare a draft EIS for the Environmental Protection Agency (EPA) where a proposed action relates to air or water quality, noise abatement or control, pesticide regulation, solid waste disposal, radiation criteria and standards, or other areas in which the EPA has jurisdiction (Turner, 2001).

Federal actions requiring environmental assessment
The trigger for determining whether an EIS is required for a proposed action is if the proposal is a "*major Federal action significantly affecting the quality of the human environment*" [NEPA Section 102(2)(C)]. The administrative decision on this procedural point is made by the Federal agency that will undertake the action. A private project can also trigger the NEPA where the private project requires Federal approval, such as a licence (Tilleman, 1995). Federal actions tend to fall within a category such as projects, programmes, plans, or policies: 40 C.F.R., Section 1508.18.

The forms of environmental assessment
Depending on whether or not a proposed action could have a significant impact on the environment, there are three forms, or levels, of assessment recognised (U.S. Environmental Protection Agency, 2007):

i. Categorical exclusions: Where the proposed action comes within one of the exclusion categories, the agency may proceed with the action, without the need to undertake some form of environmental assessment. A categorical exclusion means a category of actions which do not have a significant impact on the human environment, either as an individual action or as a cumulative impact: 40 C.F.R., Section 1508.4. Some Federal agencies have developed lists of actions which are normally categorically excluded from environmental evaluation under the EPA's regulations made under NEPA.
   If the action does not come within a categorical exclusion, the agency must prepare an environmental assessment: 40 C.F.R., Section 1501.4(b).
ii. Environmental assessment ("EA"): An environmental assessment is undertaken by a Federal agency to determine whether an action would significantly affect the environment. An environmental assessment is a "diluted" assessment relative to an EIS and its purpose is to determine

whether the more rigorous EIS is required (40 C.F.R., Section 1508.9). The agency is required to prepare a public record of its decision, together with supporting documentation. The purpose is intended to ensure that the agency has complied with its statutory procedures under the NEPA and that the agency gave the problem adequate consideration when evaluating the action (Turner, 2001). This material provides a check and balance that the agency has complied with statutory procedures in its review processes and so is relevant for avoiding a possible judicial review challenge.

If the agency decides, based on the environmental assessment undertaken, that an EIS is not required, it must then issue a "*finding of no significant impact*" ("FONSI"): 40 C.F.R., Section 1508.13.

iii. Environmental impact assessment: Where the environmental assessment determines that the consequences of a proposed Federal action may have a significant impact on the environment, an EIS must be prepared. In addition, where a Federal agency anticipates that an action may have a significant impact on the environment, or where the project is environmentally controversial, the agency may choose to prepare an EIS without undertaking an environmental assessment (U.S. Environmental Protection Agency, 2007).

Actions and significant impact

The evaluation of scientific information within the boundaries set by the statutory meaning for the term "significantly" is central to any administrative decision made in relation to the need for, as well as the form of, environmental assessment. The term significantly, as used in the NEPA, requires consideration of both the "context" and the "intensity" of the environmental impacts as defined according to their statutory meaning: 40 C.F.R., Section 1508.27. Context means that the significance of an action must be analysed in several contexts such as society as a whole (human, national), the affected region, the affected interests and the locality; both short- and long-term effects are relevant. Intensity refers to the severity of the impact.

There are nine prescribed matters prescribed in Section 1508.27 (40 C.F.R.) that "*should be considered in evaluating intensity*". They include:

- The degree to which the proposed action affects public health or safety;
- Unique characteristics of the geographic area such as proximity to historic or cultural resources, prime farmlands, wetlands or ecologically critical areas;
- The degree to which the possible effects on the human environment are highly uncertain or involve unique or unknown risks;

- The degree to which the action may establish a precedent for future actions with significant effects;
- Whether the action is related to other actions with individually insignificant but cumulatively significant impacts;
- The degree to which the action may adversely affect an endangered or threatened species or its habitat determined to be critical under the ESA; and
- Whether the action threatens a violation of Federal, State or local law or requirements imposed for the protection of the environment.

Actions and the human environment
Where either social and economic or social and natural or physical environmental effects are interrelated, the EIS must discuss all of these effects on the human environment (Environmental Law Institute, 2001). Human environment is interpreted "*comprehensively to include the natural and physical environment and the relationship of people with that environment*" (40 C.F.R., Section 1508.14).

Public participation and environmental assessment
There is an important distinction between EIA and environmental assessment that needs to be recognised. Public participation in the NEPA is recognised as a significant element for the successful implementation of every step of the EIA process and is regarded as the hallmark of the NEPA (Environmental Law Institute, 2001). The Environmental Law Institute has identified the following Regulations as key steps in public involvement under the NEPA for preparing and implementing public participation procedures (40 C.F.R., Section 1506.6 — "Public Involvement"):

- Provision of public notice of NEPA-related hearings, public meetings, and the availability of environmental documents;
- Holding or sponsoring public hearings or public meetings whenever appropriate or in accordance with statutory requirements;
- Requesting appropriate information from the public;
- Explaining procedures where interested persons can get information or status reports on environmental impact statements and other elements of the NEPA process; and
- Making environmental impact statements, the comments received and any underlying documents available to the public pursuant to the Freedom of Information Act.

However, in contrast to the EIA process, a public review of the FONSI, before the agency makes its final determination of whether an EIS is to be prepared, is usually only necessary in "*certain limited circumstances*" prescribed by legislation. Specifically, if the proposed action is, or is closely similar to, one which normally requires the preparation of an EIS, or the nature of the proposed action is one without precedent: 40 C.F.R, Section. 1501.4(e)(2).

INTERNATIONAL ENVIRONMENTAL CONVENTIONS AND DECLARATIONS

In 1992, Principle 17 of the Rio Declaration gave effect to EIA being undertaken as part of the administrative decision-making process: "*Environmental impact assessment, as a national instrument shall be undertaken for proposed activities that are likely to have a significant adverse effect on the environment and are subject to a decision of competent national authority.*"

From the time of the introduction of the NEPA in the United States in 1969, EIA has become a key part of global environmental regulation. EIA is now also prescribed in many specific treaties, e.g. the Espoo Convention on *EIA in a Transboundary Context* (1991). Strategic environmental assessment has also been adopted as an international treaty, the Kiev Protocol on *Strategic Environmental Assessment* (2003): Kiss and Shelton (2004). However, whilst there is still no global EIA treaty, Bell and McGillivray (2006) argue that EIA may have reached the status of becoming either a general principle of law or even a requirement of customary international law.

On 13 September 2007, the General Assembly of the United Nations (2007) adopted a landmark declaration, the *United Nations Declaration on the Rights of Indigenous Peoples*. It was approved after 143 Member States voted in favour; there were 11 abstentions. Four States — Australia, Canada, New Zealand and the United States — voted against the text. The Declaration is non-binding and promotes, amongst other things, the full and effective participation by Indigenous peoples in all matters that concern them, and their right to remain distinct and to pursue their own visions of economic and social development.

LEGAL RIGHTS AS A SOURCE OF KNOWLEDGE POWER FOR NEGOTIATING EIA CONFLICTS

Procedural obligations imposed by EIA legislation arise at various stages of the EIA process — commencing from the time a decision is made to determine whether an EIS is needed. Procedural obligations then continue through to obligations for public participation, review and assessment procedures and, finally, project authorisation. Compliance with procedural obligations is fundamental if a Government agency is to avoid a judicial review challenge for administrative decisions made throughout the EIA process. An error of law will arise, and so be a ground for a judicial review, if prescribed procedures under the EIA statute were not adhered to in the making of an administrative decision. The ever-present potential threat of litigation is the reality, especially as the experience in the United States illustrates where excessive legalism is associated with compliance to procedural obligations in the NEPA. Consider the following examples of procedural obligations confronting a Government agency in the administrative decisions it is required to make throughout the EIA process:

i. Ensuring the consideration of all matters that must be taken into account, as a legal obligation, in determining whether a proposed action or project is likely to have a significant impact on the environment;
ii. Information provided to the public of the environmental impacts of the proposal and the consequences to the public inherent in the carrying out, or not carrying out, of the proposal is part of the EIA process. Inaccurate, incomplete and inaccessible information may make full access to information in the EIA process problematic (Sinclair and Doelle, 2003);
iii. Common law in the UK has established clear legal principles for procedural fairness and legitimate expectation that require any consultation process to be full and fair to enable those consulted to make an intelligent response. Legitimate expectation is a long-held accepted principle of the law of procedural fairness in Australian public law. Consequently, principles developed in the UK for legitimate expectation and information disclosure, as part of the community consultation process, would justify consideration in Australia as a persuasive precedent; and
iv. In the United States, there is a procedural obligation for the purpose of an EIS to provide a "*full and fair discussion of significant environmental impacts*" (40 C.F.R., Section 1502.1). Therefore, information disclosed in the public participation process, and for which public comment is sought, places a special onus on the Government agency to comply with this obligation. However, in order for the UK public law principles to apply in the United States, the law relating to procedural due process would need to have developed equivalent principles for legitimate expectation in terms of a full and fair consultation process and information disclosure when the environment is in issue.

In terms of standing, there are wide statutory standing requirements prescribed under the EPBC Act in Australia. Similarly, in the United Kingdom public law principles result in standing being widely available. In terms of consultation and EIA, it is relevant to note that in the UK case, *R v Swale Borough Council and Medway Ports Authority ex parte The Royal Society for the Protection of Birds* [1991] JPL 39, a community group was granted standing in their judicial review challenge for a situation in which a planning permission was granted but without environmental assessment being required or carried out. The court found the community group to have a legitimate expectation for being consulted as they had been given a written assurance by the planning authority that they would be consulted (Wolf and White, 1997). In the United States, the NEPA has no applicable provisions for citizen suits (Environmental Law Institute, 2002). Therefore, the tests for standing for judicial review in the Federal Court will apply.

The challenge therefore becomes whether "excessive legalism" related to procedural obligations can be turned into a positive. Can EIA conflicts be managed and resolved by ADR and negotiation, rather than through litigation and the courts? Can knowledge of legal rights provide the parties with an adequate basis of power to participate effectively in negotiations related to an EIA conflict? Given the legal rights that relate to procedural obligations and the EIA process, together with the accessibility of the public to the courts through standing, an arguable case could be made that it would be prudent for a Government agency to consider alternative participatory approaches that involve sharing of power to manage and resolve conflicts that may arise in the EIA process. The benefits for the Government agency would be that not only would delays arising from continuous litigation be offset, but also the EIA process would become an effective aid for administrative decision-making.

POWER, INFORMATION DISCLOSURE AND THE EIA PROCESS

Information disclosure is a central element of the EIA process as EIA legislation provides for strategic information being made available for public comment, extending to the systematic appraisal of the positive and negative effects on the environment that are possible, or likely to arise, from a proposed action. However, public disclosure of environmental information in the EIA process and its review and assessment are not without some difficulty for administrative decision-making. In particular, how to reconcile the need for administrative decisions to be based on a full and fair disclosure of all relevant information to the public, with public concerns over scientific uncertainty and the accuracy of predicted environmental impacts. Limitations in promoting and mobilising the use of Indigenous traditional knowledge in the EIA process add a further dimension to the problem for decision-making.

(a) Limitations and Uncertainty in Scientific Information

Limitations in the capability of science to deliver precision in predicting the severity of potential adverse environmental impacts, as well as scientific uncertainty in the available information, create difficulty for the administrative decision-maker. A major procedural step in the United States to address scientific uncertainty in the EIA process was the introduction of a Regulation for "*Incomplete or Unavailable Information*": 40 C.F.R., Section 1502.22. Where "*information relevant to evaluating reasonably foreseeable significant adverse impacts*" in an EIS cannot be obtained because (i) the overall costs of obtaining it are exorbitant or (ii) the means to obtain it are not known, this Regulation requires the agency to include, within the EIS:

- A statement that such information is incomplete or unavailable;
- The relevance of this information for evaluating reasonably foreseeable significant adverse impacts;
- A summary of relevant existing credible scientific evidence on such impacts; and
- The agency's evaluation of such impacts based upon generally accepted scientific theories and research methods.

Mattix and Becker (2002) contend that this Regulation aims to set aside any belief that more science is required, when the effects of those consequences have not been fully analysed. There is no equivalent legislation that addresses limitations in scientific information in Australia or the UK. In the analysis of United States case law under 40 C.F.R., Section 1502.22 undertaken by Mattix and Becker (2002), the following legal principles emerge:

- A distinction can be made for scientific uncertainty due to "*unattainable or unknowable science*" and a clear omission on the part of an agency to discuss the full environmental effects of a proposed action: *Blue Mountains Biodiversity Project v Blackwood* 161 F.3d 1208 (9th Cir. 1998);
- Scientific information, e.g. a theory or principle, which is too speculative is not "*reasonably foreseeable*" and will not trigger the threshold question under the Regulation as to whether there is a

"*reasonably foreseeable significant adverse effect*": *Sierra Club v Marita* 47 F.3d 606 (7th Cir. 1995); and
- In *Colorado Environmental Coalition v Dombeck* 185 F.3d 1162 (10th Cir. 1999), the court concluded that there had been a failure "*to show how additional site-specific* [ecological] *data* [are] *essential to reasoned decision-making*". The ecological data sought did not exist and were not obtainable. Furthermore, the agency had used the best available data and applied the accepted scientific methodology in its assessment of the relevant adverse ecological impacts.

(b) Cumulative Impacts

The assessment of cumulative impacts has long been recognised as a potential limitation of the EIA process, notwithstanding that it may be critical for the administrative decision to be made. Cumulative impacts may also be a source of conflict where competing interests for land use intersect. Thornton (1991) argues that it could be expected that developers would seek to have the analysis of cumulative impacts limited sharply, whereas environmentalists would aim to broaden the scope of the analysis. The term "cumulative impacts" has not been defined in EIA legislation in Australia or the UK/EU. However, in the United States, "cumulative impacts" have been defined as:

> "[t]he impact on the environment which results from the incremental impact of the action when added to other past, present, and reasonably foreseeable future actions regardless of what agency (Federal or non-Federal) or person undertakes such other actions. Cumulative impacts can result from individually minor but collectively significant actions taking place over a period of time" (40 C.F.R., Section 1508.7).

The uncertainties for administrative decision-making for assessing cumulative impacts relate to time, or temporal bounds, and space, or spatial bounds (Sample, 1991). Consider the following scenario: a proposal for a new coal-fired power station is subject to an EIS. The carbon dioxide emissions from the proposal may be minor, but the impact of incremental impacts arising collectively from other coal-fired power stations may have the potential to result in a reasonably foreseeable adverse cumulative impact for the greenhouse effect from the proposal. In such a situation, some of the problems confronting an administrative decision-maker in deciding whether a future action is reasonably foreseeable rather than speculative, as a cumulative impact, have been highlighted by Sample (1991):

- In terms of time, how far in the future should the decision-maker consider similar and connected actions and their cumulative effects?

- In addition, should safe use thresholds be approached or exceeded, there is a need to consider the length of time that the impacts would persist. The extent of recovery for the environment would need to also be considered as a function of time; and
- Spatial bounds are also problematic and a key issue is whether boundaries should be established on an administrative basis (e.g. local or Federal authority, town, national park) or some biophysical basis (e.g. a watershed or ecosystem).

Given these areas of scientific uncertainty, it is not surprising that some legal commentators on reviewing case law and cumulative impacts in the United States have concluded that there have been inconsistencies in the application of the legislative (40 C.F.R., Section 1508.7) meaning of cumulative impacts (Thornton, 1991). Sample (1991) contends that perhaps the most compelling cumulative impact argument would be based on the potential effects of several actions on one particular resource that can be specified — i.e. a "target resource". The target resource could be a wildlife species, its habitat, a form of recreation or some prescribed environmental value. Sample refers to the case *Natural Resources Defence Council Inc. v Hodel* 865 F.2d 288 (D.C. Cir. 1988), where a challenge was made to several off-shore drilling proposals extending from Alaska to southern California. The court found the EIS to be flawed in failing to consider cumulative impacts: specifically, the target resources of salmon and whales that migrated past the widely separated locations of activity.

In *Queensland Conservation Council Inc. v Minister for the Environment and Heritage* [2003] FCA 1463, Keifel J of the Federal Court of Australia (as she then was) concluded that possible environmental impacts which lay in the realms of speculation were excluded from further consideration as cumulative impacts in the EIS. In this case, a judicial review challenge was successful. The EIS for the dam proposal failed to consider likely cumulative impacts on the Great Barrier Reef World Heritage Area arising from irrigation downstream from the dam through the use of water and its potential to carry and release nutrients and chemicals.

(c) Strategic Environmental Assessment and Cumulative Impacts

Given the problematic nature of identifying and assessing cumulative environmental impacts, it is particularly relevant that strategic environmental assessment ("SEA") represents an accepted scientific method for evaluating cumulative impacts. SEA is a formal, systematic process that analyses and addresses the environmental effects of policies, plans and programmes, at the earliest appropriate stage of decision-making. SEA applies primarily to development-related initiatives likely to have significant environmental effects, e.g. collectively through spatial or land use change (Abaza *et al.*, 2004). Like EIA, it is an aid to administrative decision-making. The advantages of SEA, relative to EIA (Marsden, 2002; Abaza *et al.*, 2004), include:

i. Cumulative impacts are not always addressed in the EIA process where the primary focus is the individual project;
ii. SEA can provide an early warning of cumulative impacts — including those arising from smaller-scale projects that individually would not significantly affect the environment. EIA, in contrast, may miss regional impacts and cumulative impacts from multiple projects;
iii. Multiple projects may be more effectively addressed by SEA as a broader range of environmental effects and consequences are assessed;
iv. Larger-scale and/or longer timeframes for assessing environmental effects and consequences may be taken into account; and
v. By identifying what forms of development are environmentally sound, SEA contributes to sustainable development.

In Australia, the Commonwealth Environment Minister may agree to conduct a strategic assessment of actions that may be carried out under a proposed policy, programme or plan: EPBC Act, Section 146. This enables the early assessment of the cumulative impacts of relevant actions under that policy, programme or plan. The outcomes of a strategic assessment may also be taken into account in deciding the appropriate form of environmental assessment that is required for a particular action (Department of the Environment and Water Resources, 2007b).

In the UK, the Directive on SEA (Directive 2001/42/EC *on the assessment of the effects of certain plans and programmes on the environment*) was implemented into domestic law in 2004: *Assessment of Plans and Programmes Regulations 2004*. The Kiev Protocol on *Strategic Environmental Assessment* (2003) follows the EU Directive on SEA (Kiss and Shelton, 2004).

In the United States, SEA was a requirement under the NEPA. However, audits on SEA have concluded that it is ineffective as most of the information under the process is made available after the decision has been taken and so too late for crucial matters (Clark, 2000).

PARTICIPATORY DECISION-MAKING, SHARING POWER AND THE EIA PROCESS

Participatory approaches in the EIA process need to reflect the level of power that the parties have to influence, share or control the administrative decision-making process. Sinclair and Doelle (2003) consider that public participation in the EIA process varies along a continuum. The extremes vary from simply ensuring that there is transparency around the administrative decision, to the more expansive goal of ensuring active public participation in the administrative decision to be made. In addition, for public participation to facilitate the administrative decision-making process in a meaningful way, public inputs are required throughout the EIA process, especially in cases where scientific uncertainty and information conflicts prevail.

(a) Panel Reviews

One approach to facilitate the participatory role of the public in the EIA process is through the use of panel reviews. The rationale for panel reviews is that they provide an increased role for the public to participate personally, fully and effectively in the environmental assessment process for project proposals that affect local communities (Rutherford and Campbell, 2004). Panel reviews have been mandated in Canada: *Canadian Environmental Assessment Act, S.C. 1992*. Under this statute, a panel review guarantees opportunities for public comment during the assessment process rather than in the late stages of the project and assessment cycle (Sinclair and Doelle, 2003). Public influence over project decisions can be both direct and indirect. A direct influence arises when the public persuade the proponents to make changes. The public may also indirectly influence the administrative decision-maker through recommendations made to the panel. The panel report and recommendations are not binding, but are provided to the Environment Minister and the appropriate Federal agency for a decision (Rutherford and Campbell, 2004).

Rutherford and Campbell refer to a review of Boyd (2003) that revealed that of the 46 projects conducted from 1995 to 2000 that were subject to comprehensive Federal environmental assessments in Canada; only 10 projects were reviewed by panels that held public hearings. Rutherford and Campbell express concern that the trend indicates that panels are not utilised enough — notwithstanding that they contend panel reviews "*tend to be the best process available to facilitate the actual exchange of ideas on whether and how a project should be allowed to proceed*". However, a panel review, unlike ADR and negotiation, is limited in that the process offers little or no direct opportunity for the public to share power with the Government agency in the decision-making process (Fiorino, 1990).

(b) Negotiation

The limitation in power-sharing in a panel review can be offset, whilst retaining all of the benefits of a panel review, where ADR and interest-based negotiation is used as the approach for participatory decision-making. On a case-by-case basis, ADR can be effectively applied at discrete stages of the EIA process to manage and to resolve information conflicts. However, the situation in Australia, the UK and the United States is that the use of ADR is not mandated in EIA legislation as part of any public participatory procedures. In Canada, however, national legislation, the *Canadian Environmental Assessment Act, S.C. 1992* (at Sections 29–32), prescribes mediation as a participatory approach for the EIA process; any issue relating to an environmental assessment by a review panel may be referred to mediation. However, the statute does not provide details on mandated guidelines for procedures and processes.

Consequently, the central issue becomes whether a Government agency in Australia, the UK or the United States should incorporate ADR procedures as part of the EIA process, either through legislation or policy guidelines. This issue needs to be considered in terms of the following benefits that ADR and a consensus-building approach would bring for administrative decision-making in the EIA process:

i. ADR and interest-based negotiation procedures represent an objective approach for offsetting the inherent limitations in the EIA process associated with information conflicts;
ii. These procedures would avoid any perception that participation in the EIA process manipulates rather than involves the public. For example, criticism that participation in the EIA process through public comment on the draft EIS has limited value because, at this stage, the proponent and the Government agency may have become effectively locked into various aspects of the proposal (Bates, 2006);
iii. Any concern that the EIS process was primarily a review on how to minimise environmental effects, rather than an assessment of whether the proposal is desirable or warranted, would be offset (Formby, 1987); and
iv. Whilst delays may be encountered in adopting ADR and interest-based negotiation, in all likelihood there would be fewer delays encountered compared with litigation and the courts being used to challenge decisions made in the EIA process.

Given that the EIA process is an aid to decision-making and not the decision end-point in itself, it is argued that the use of ADR and interest-based negotiation to manage and resolve information conflicts in the EIA process has advantages over existing approaches for consultation and participation. Not only do the parties share power with the Government agency in administrative decision-making but there are further benefits for all parties arising from ownership in outcomes reached by consensus.

(c) Public Notice and Comment

The alternative to panel reviews or negotiation is to persist with the existing participatory procedures for EIA, in which a Government agency is only required to take into account public comment in its administrative decision-making process. The use of litigation and the courts to challenge administrative decisions made in response to procedural obligations arising from the EIA process would be ever-present. Litigation may result in the purpose of the EIA process, as a decision-making aid, being diminished. In the extreme case, litigation may even lead to a Government agency relying on forms of environmental assessment, other than EIA, that have limited public review inputs — a self-defeating outcome given that public participation is a cornerstone of the EIA process.

FINDING SOLUTIONS FOR MANAGING AND RESOLVING INFORMATION CONFLICTS IN THE EIA PROCESS

Abaza *et al.* (2004) believe that there are three main trends for participatory approaches in the EIA place that are likely to have increasing influence, but acknowledge there will be variations in their use in time and place:

- An increasing role for participatory approaches;
- The use of mediation in conflict resolution; and
- The use of Indigenous traditional knowledge.

Any evaluation for a change in focus to increase the role for participatory approaches in the EIA process should consider the following characteristics as linked — rather than discrete — factors that affect the integrity of administrative decision-making in the EIA process:

i. All administrative decisions arising from procedural obligations imposed by each stage of the EIA process could be the subject of a judicial review challenge — and not only the final project authorisation decision (Weston, 2002);

ii. Consultation and participation are cornerstones of EIA legislation. The EIA process provides for a wide range of procedures to involve the public;

iii. EIA legislation prescribes procedures for public participation, but there is no statutory obligation imposed on the administrative decision-maker to be bound by the outcome of public participation;

iv. The administrative decision-maker has a broad discretion on what to do with the public comments received. Sinclair and Doelle (2003) state that a problem for trust in the EIA process will arise where there is a perception that the goal of public participation is to manipulate rather than to involve the public;

v. The requirements for procedural fairness for full and fair disclosure of information in the EIA process need to be understood as a potential source for a judicial review challenge;

vi. Depending on the form the assessment of environmental impacts takes, some forms of environmental assessment may not have the same scope for public participation compared with the EIA process, e.g. assessment on preliminary documentation in Australia and an environmental assessment in the United States;

vii. Kubasek and Silverman (2005) refer to a 2003 study published by the EPA and note that, in the United States, the number of EISs published by Federal agencies steadily declined from around 2000 reports in 1973, levelling off to about 500 reports, annually, from 1990–2001. In contrast, environmental assessments became increasingly popular. Kubasek and Silverman refer to a survey of Federal agencies conducted by the Council of Environmental Quality in 1993 that indicated that approximately 50,000 environmental assessments were being prepared annually. The authors refer to a concern raised by environmental groups who "*fear that certain EA's are filed to avoid public involvement*";

viii. Information conflicts in the EIA process represent a potential and significant source of conflict for ongoing litigation. The end result, in all likelihood, would be to reduce the effectiveness of the EIA process to assist administrative decision-making; in the extreme case, it could defeat the objectives of the EIA legislation; and

ix. Acceptance that information conflicts are problematic and inevitable in the EIA process requires some understanding of the grounds that may lead to a judicial review challenge in relation to information conflicts.

Abaza *et al.* (2004) state that there is a "*middle ground*" between consultation and participation in which shared involvement and responsibility for decision-making in the EIA process could prevail and so benefit the parties and the likely success of the proposal. Abaza *et al.* call for innovatory approaches for shared involvement and responsibility, e.g. by mediation and consensus-building.

ADR and interest-based negotiation involves collaborative fact-finding and problem-solving as well as consensus-building, and so has greater power, relative to panel reviews, to influence administrative decisions made in response to procedural obligations over the entire EIA process. ADR and interest-based negotiation can effectively facilitate information conflicts — the most complex and problematic area of the EIA process. The need is for flexibility, first, by using ADR to manage information conflicts in the earlier stages of the EIA process. Should information conflicts persist in later stages, then the focus of ADR would be to resolve information conflicts in a procedurally fair manner. Accordingly, ADR and interest-based negotiation could be incorporated in the EIA process with a focus on managing and resolving information conflicts at sequential stages of the EIA process. Power would be shared between the Government agency and parties having a recognised or representative interest in the administrative decision to be made.

i. *Screening stage: management of information conflicts*: At the commencement of the EIA process, issues associated with scientific uncertainty, the potential inaccuracy of predicted environmental impacts and incomplete and inaccessible information may be difficult to reconcile with the need for procedural fairness in the EIA process through a full and fair disclosure and access to information. The use of data (information) mediation (Susskind and Weinstein, 1980) would enable a common database, based on the best available scientific information, to be identified at the outset.

    Next, an administrative decision must be made whether the proposed action or project is likely to significantly affect the environment in order to determine whether an EIS (or environmental statement in the UK) is required. The EIA legislation in the UK and the United States, and policy guidelines in Australia, specify the matters which characterise the term "significant". In addition, cumulative impacts may be central to the EIA process but pose difficulty for the administrative decision-maker. Their assessment may be confounded by uncertainty in identifying whether a future action or project is reasonably foreseeable, in terms of time and space, as an adverse cumulative impact.

    Both these issues, reflecting different sources of information conflict, could be effectively addressed using the ADR-based model, the scientific round table, thereby ensuring the integrity and consistency of the administrative decision-making process. The scientific round table would involve scientific experts, nominated by and representing the parties, and would be an effective means for managing these sources of scientific information conflict

ii. *Review and assessment stage: management of information conflicts*: Where divergent scientific opinion existed on conclusions contained in the draft EIS (or environmental statement in the UK), the scientific round table would be an effective procedure for facilitating agreement on disputed scientific and technical issues. Any such agreement would need to be consistent with the common database as identified in the data mediation.

iii. *Project authorisation stage: resolution of conflict through interest-based negotiation*: Draft recommendations made in relation to project approval, including conditions imposed to mitigate or to manage potential adverse environmental impacts, may create controversy and further conflict. Abaza *et al.* (2004) point out that mediation and consensus-building could also encompass shared responsibility for the implementation of the proposed action or project. Approaches could be used for the parties to be collaboratively involved in preparing the environmental management plan and the ecological monitoring programme.

    For this stage, a summary of agreed outcomes for the specific issues in dispute would be made available to the parties from the scientific round table. The ADR process used would be "evaluative mediation". Using an interest-based approach, the parties would seek solutions in the form of a negotiated agreement to resolve conflict related to the action or project.

TRADITIONAL INDIGENOUS KNOWLEDGE AND THE EIA PROCESS

The different perspective that Indigenous traditional knowledge has for the causes and consequences of environmental change, natural resource management and social interactions needs to be reflected in the EIA process; the EIA process must accommodate Indigenous knowledge in a meaningful manner within the context of the dominant knowledge system (Abaza *et al.*, 2004). As these authors point out, any failure to understand and to effectively apply Indigenous knowledge in the administrative decision-making process will lead not only to limitations in the available information for the administrative decision-making process, but also create potential uncertainty for identifying and predicting impacts. In the extreme case, conflict and litigation will arise. An example of how such a problem can arise is illustrated in the expert evidence given by an anthropologist in *Re Zen Pearls and the Great Barrier Reef Marine Park Authority & The Manbarra People* [2004] AATA 267, an Australian case under the GBRMP Act. The case involved a pearl oyster mariculture in the Great Barrier Reef Marine Park:

> " ... the underlying problem with the pearl farm being there is ... is the fact that it's there without the [Indigenous] community feeling they've been given their consent for it to be there. That's what's made people angry ... well, I think that's quite logical because if the Marine Park [Authority] can simply decide to allow ... an infrastructure to be built within the traditional waters of the [indigenous] people, and ... the Elders do not give their consent, every time it does that it undermines the system of traditional law."

In the United States, under the NEPA, where the effects of a proposed action are on a reservation, a Tribe by agreement with the lead Government agency may become a "*cooperating agency*": 40 C.F.R., Section 1508.5. The role of a cooperating agency includes participating in the scoping process and, at the request of the lead agency, responsibility for developing information and preparing environmental analyses, including portions of the EIS: 40 C.F.R., Section 1501.6(a).

Indigenous people in Australia have significant legal rights incorporated in the statutory objectives of the EPBC Act that relate to recognising the role of Indigenous people in the conservation and ecologically sustainable use of Australia's biodiversity [Section 3(1)(f)] as well as promoting the use of their knowledge of biodiversity [Section 3(1)(g)]. These matters are relevant for administrative decision-making in the EIA process, for the appropriate proposed action or project, as considerations that must be taken into account as a legal obligation.

Abaza *et al.* (2004) raise two possibilities for accommodating Indigenous traditional knowledge in the EIA process that warrant consideration:

i. *The production of multiple impact scenarios*: One scenario would be based on the "Western" scientific model, the other based on Indigenous traditional knowledge. The different perspectives of the same proposal in terms of its consequences and effects on the environment would enhance the understanding of the needs and concerns of all interests in relation to the proposal. The benefit of such a procedure is that it would maximise the information available to the administrative decision-maker. In addition, it would ensure that the decision-maker was fully aware of the consequences to the community inherent in the carrying out, or not carrying out, of the proposal; and

ii. *Indigenous people to produce some EISs*: This approach for inputs into administrative decisions by Indigenous people enables a proposed development to be considered in terms of its consistency with Indigenous aspirations, cultural issues, community development and the preservation of ecologies. Alternatively, in specific circumstances, Indigenous knowledge becomes the environmental basis for administrative decisions.

These possibilities are not mandated, as yet, in EIA legislation. However, as a cornerstone of the EIA process is full information disclosure, the need for innovative, participatory approaches for managing and resolving information conflicts in the EIA process is equally justified for Indigenous traditional knowledge as it is for the "Western" scientific model.

LITIGATION V NEGOTIATION AND ENVIRONMENTAL CONFLICTS

Camacho (2005) states that courts are ill-equipped to evaluate and determine the most appropriate outcome involving competing parties affected by a typical land use conflict, in particular the requirement large-scale projects have for information to be obtained and evaluated from diverse sources, e.g. scientific, social and economic information. Courts also have limitations relating to the trading of preferences together with the need to balance public policy concerns that affect multiple, subjective and disparate interests.

In contrast to litigation, ADR and interest-based negotiation enables information conflicts in the EIA process to be managed and resolved collaboratively in a problem-solving process. This advantage flows into implementation of outcomes and solutions agreed to through consensus and negotiation, resulting in a sense of ownership by the parties. In addition, a participatory approach based on consensus and negotiation facilitates the aim of an EIS to be an aid to administrative decision-making being achieved. This participatory approach to decision-making has a further advantage in that information conflicts may be effectively resolved at the project authorisation stage. Unresolved information conflicts at this stage that have the potential to erupt into litigation at a later date would be avoided. Finally, ADR and interest-based negotiation has significant advantages over litigation and the courts for finding solutions for sustainable development conflicts in terms of considering scientific, social and economic information and the need to balance public policy concerns that affect multiple parties holding competing interests.

7. Risk, precaution and the environment: biotechnology

"... the dust of battle over what amounts to an acceptable risk in relation to the release of a genetically modified organism will almost certainly obscure the truth."

Sampson (2003)

"The legal system has no special competence to assess and compare public risks, and the legal process is not designed or equipped to conduct the broad-ranging aggregative inquiries on which sensible public-risk choices are built."

Plater *et al.* (2004)

"No matter how much the specialists sneer at an "irrational" and "ignorant" public, lay judgements about possible dangers are equally as important as scientific or technical analysis."

O'Riordan (1989b)

"... a constant application of extreme precaution in the absence of risk-based data and decision making is anti-scientific and even verges on superstition."

Manson (2004)

"[A mathematical model] enables us to weave a tapestry in which threads of fact and fiction are combined to portray an impression of [the environment]. The result is a work of art, sometimes good, sometimes bad, but almost always giving the creator a feeling of euphoria. This euphoria often lives on for some time unless we make the mistake [of validating the model output]."

Passioura (1973)

INTRODUCTION

Genetically modified organisms are new products arising from the technological manipulation of the genetic material of a living organism through means other than a natural recombination. Biotechnology is an emerging industry in an era of globalization and has been defined as "*the use of biological processes for the development of products such as foods, enzymes, drugs and vaccines through the modification, alteration and manipulation of genetic material of plants, animals and microbic systems to derive or to produce new products and technologies*" (Vogt and Parish, 2001). Genetic modification has applications to biotechnology in the areas of agriculture, forestry, medicine, environmental remediation and industry. However, biotechnology continues to be a controversial topic. Proponents of biotechnology claim that genetic modifications can be less harsh on the environment, e.g. through a reduced use of agricultural chemicals as well as contributing to environmentally sustainable development and greater food production. Critics, on the other hand, express concerns about the long-term risks of the escape of genetically modified organisms ("GMOs") into the environment through cross-fertilisation. Additionally, there is a fear that introduced genes in "supercrops" and novelty foods could prove allergenic or harmful to human health (Vogt and Parish, 2001). The opportunities created by biotechnology applications to benefit humankind are both broad and persuasive. Nevertheless, the opportunity needs to be counterbalanced against risk.

The release of GMOs into the environment persists as a source of global concern and needs to be considered in the context of GMOs representing a "hazard" as well as a "risk". A hazard refers to a source of potential harm to the environment and to human health. A risk refers to a chance or probability that environmental harm or harm to human health will actually occur. Techniques for evaluating hazards and quantifying risks developed as an engineering discipline, and for insurance purposes, facilitated by two influential reports by the Royal Society in Britain and the United States National Research Council. Environmental agencies, primarily the United States Environmental Protection Agency (EPA), adopted risk assessment and risk management as an objective method to set standards, set priorities and to assist administrative decision-making (Beer and Ziolkowski, 1995). Following on from the engineering approach to quantified risk assessment, health risk assessment emerged as a technique for assessing health risk, particularly cancer risk. Health risk assessment has been an accepted scientific method for over two decades, notwithstanding that there has been considerable debate over its applications. A subsequent extension of health risk assessment is environmental (sometimes referred to as ecological) risk assessment that is used to assess risk to the environment. Central to risk management is the concept of acceptable level of risk — a concept that is both subjective and debatable and which requires a balancing of scientific knowledge with social, economic and policy considerations.

The categories of risk associated with biotechnology extend beyond "scientific risk" — such as human health and environmental harm — to include risk perception (Pandya, 2004). Risk perception involves people's feelings, beliefs, attitudes and judgements and is at the core of public opposition to a range of potentially hazardous technologies and concerns (Otway and Winterfeldt, 1982). Barnes (2001) provides a detailed analysis of risk and risk communication from a social science perspective. Factors are identified which when present, lead to public disbelief and concern being expressed — even though reassurances are made by Government agencies that there is minimal risk to life or health associated with a particular hazard or technology. The following factors are accepted as important in understanding public perceptions over risk and safety:

i. A lack of public trust in the Government agencies involved in the control of hazards and the regulatory control of risk (Otway and Sims, 1987; Slovic, 1993);
ii. A failure to disclose key pieces of information; and
iii. Inaccuracies in the communication process (Wynne, 1989).

In terms of biotechnology and the genetic manipulation of plants, the EU–US Biotechnology Consultative Forum (2000) concluded that, although it may be unattainable to prove in advance that each and every action is risk free, the prevention of future mistakes together with ensuring public trust in the integrity of administrative decision-making require precaution. The advent of the precautionary principle extends the applications of the concept of risk assessment and biotechnology. The focus of the precautionary principle is to guide environmental decision-making, when scientific uncertainty exists, in order to avoid, wherever practicable, serious or irreversible environmental damage. One element for the application of the precautionary principle is an assessment of the risk-weighted consequences of various options for environmental protection, as measures to manage risk. Accordingly, the use of risk assessment methodology is central to the application of the precautionary principle — not only for maintaining the integrity of the administrative decision-making process, but also to ensure procedural fairness in the decision end-point for parties holding competing interests towards development and the environment (Christie, 1993d).

The assessment, management and communication of risk are all key factors central to decision-making for environmental risk (Beer and Ziolkowski, 1995). In the context of biotechnology and the release of GMOs into the environment, the administrative decision-making process used to assess scientific risk and public perceptions of risk will be dependent on the participatory process used by a Government agency. The EU–US Biotechnology Consultative Forum (2000) considers that the administrative decision-making process should ensure that the public have confidence in the

Government agency, that it is transparent, involves all interested parties and for trust to prevail.

BIOTECHNOLOGY AND SUSTAINABLE DEVELOPMENT

There would be little dispute that achieving sustainable agriculture in developed and developing countries is a major, contemporary global need. The World Summit on Sustainable Development at Johannesburg, South Africa, in 2002, listed "*social development*" as one of the three pillars of sustainable development. The *Johannesburg Declaration on Sustainable Development* (at paragraph 19) specified "*hunger and malnutrition*" as one of the worldwide conditions that represents a severe threat to sustainable development (Kiss and Shelton, 2004). Earlier, a similar conclusion had been made by the EU–US Biotechnology Consultative Forum (2000). In recognising that concerns associated with biotechnology related to sustainable agriculture, the Consultative Forum acknowledged the role of biotechnology for achieving the objectives of sustainability and global hunger.

However, there remains a concern about the extent that biotechnology can be integrated into sustainable development. Redgwell (2006) recognises two such concerns. First, that integration needs to achieve an equitable distribution of the benefits and costs of biotechnology — not only within existing generations, but also between generations. In addition, given the potential risks to biodiversity arising from the release of GMOs, a responsibility to future generations for the conservation and sustainable use of biodiversity also exists.

LEGISLATIVE APPROACHES FOR THE REGULATORY CONTROL OF BIOTECHNOLOGY

The focus of this chapter is on the deliberate release of GMOs into the environment and their potential threat of risk to the conservation and sustainable use of biodiversity, as well as risks for human health. It should be recognised that regulatory control of biotechnology also extends to contained use of GMOs in laboratory experimentation through to food safety.

The rationale for regulatory control in Australia, the UK and the United States is to acknowledge that an identified risk to the environment, or to human health, may be managed and reduced to some acceptable level of risk, rather than to demand a zero risk situation. There are characteristic differences between the extent of public participation in the assessment and management of risk as well as the adoption of the precautionary principle. A precautionary approach has been adopted in Australia and the UK based on anticipating environmental hazards that have not been detected but which could occur. In contrast, the United States has adopted a preventative approach that attempts to minimise environmental harm where harm has been scientifically established (Patterson and Josling, 2001).

(a) Australia

The *Gene Technology Act 2000 (Cth)* (the "GT Act") commenced on 21 June 2001 and focuses on the perceived problems associated with introducing GMOs into the Australian environment. The intention for the GT Act is to form a component of a "*nationally consistent scheme*" with the States, for the regulatory control of GMOs (s. 5). The GT Act establishes the independent position of the Gene Technology Regulator (the "Regulator") and the Federal agency, the Office of the Gene Technology Regulator. The GT Act takes a risk management approach to regulatory control, as reflected in its object (at s. 3): "*to protect the health and safety of people, and to protect the environment, by identifying risks posed by or as a result of gene technology, and by managing those risks through regulating certain dealings with GMOs.*"

Central to the object of the legislation is that the risks imposed by gene technology can be managed rather than for the technology to be rejected. There is a statutory obligation under the GT Act to prepare a risk assessment and risk management plan (s. 51). The GT Act [s. 4(aa)] gives effect to the precautionary principle as part of the regulatory framework for achieving the object of the legislation: " ... *where there are threats of serious or irreversible environmental damage, a lack of full scientific certainty should not be used as a reason for postponing cost-effective measures to prevent environmental degradation.*"

Licensing and regulatory control

The administrative decision to issue a licence requires the Regulator to be "*satisfied that any risks posed by the dealings proposed to be authorised by the licence can be managed in such a way as to protect the health and safety of people and the environment*" (GT Act, s. 56). "*Deal with*", in relation to a GMO, is defined (s. 10) to include "*conducting experiments*", "*grow, raise or culture*", "*make develop, produce or manufacture*" and "*the use, transport and disposal of*" GMOs. All dealings are prohibited under the GT Act and so require a licence, unless the dealing is subject to a number of qualifications, e.g. dealings that are either exempt or a notifiable low risk dealing [s. 32(1)(c)(d)]. The statutory framework and key procedural obligations under the GT Act and the *Gene Technology Regulations 2001 (Cth)*, upon which an administrative decision is made by the Regulator to issue a licence — with or without conditions — or to refuse a licence may be summarised as follows:

i. Where the Regulator is satisfied that dealing with a GMO may pose a "significant" risk to the health and safety of people or the environment, public notice of the application must be published inviting written submissions on whether the licence should be issued: GT Act, s.49(1);
ii. The matters that the Regulator is required to take into account, as a legal obligation, in determining whether dealings proposed to be authorised by a licence may pose significant risks to health and safety and the environment are prescribed in section 49(2);
iii. The term "significant" is the trigger for the administrative decision-making process to prepare a risk assessment and risk management plan before the issue of a licence (GT Act, s. 50). However, there is no statutory meaning provided for this term to ensure consistency and integrity in the administrative decision-making process for objectively evaluating the matters prescribed in section 49(2). Settled common law for EIA in Australia, for example, defines the term "significant" in relation to both "context" and "intensity". The legal uncertainty created for failing to define this term is not only a problem for the administrative decision-maker, but also a potential source for a judicial review challenge, e.g. a declaration by the court on the legal meaning and application of the term "significant";
iv. The Regulator must prepare a risk assessment and risk management plan whether or not a notice was required to be published in relation to the application [GT Act, s. 50(2)];
v. The Regulator must seek advice on matters relevant to the preparation of the risk assessment and management plan, but only from the following interests specified [GT Act, s. 50(3)]: the States and Territories of Australia, the Gene Technology Advisory Committee, prescribed Government agencies and any local authority considered appropriate by the Regulator; and
vi. Prior to the administrative decision being made, the Regulator must publish a notice seeking written submissions from the public on the risk assessment and management plan (GT Act, s. 52). The same Government interests who had earlier given advice on matters relevant to the preparation of the risk assessment and management plan may also provide submissions on the risk assessment and management plan at the same time that public submissions are sought.

Policy considerations and the Gene Technology Act

The Regulator has the discretion to consider a number of policy matters under the GT Act. The policy matters include:

- The risk assessment and risk management plan: GT Act, ss. 56(2)(a,b);
- Any submissions received on the risk assessment and risk management plan: GT Act, ss. 56(2)(c); and
- Any policy guidelines issued by the Ministerial Council relating to risks and ways to manage risks: GT Act, ss. 56(2)(d).

A key policy providing a methodology for assessing and managing risks is the *Risk Analysis Framework for Licence Applications to the Office of the Gene Technology Regulator (2005)*, prepared by the Office of the Gene Technology Regulator [the "*Australian Risk Analysis Policy (2005)*"]. The administrative decision by the Regulator must also be consistent with any policy principles issued by the Ministerial Council: GT Act, ss. 21, 51.

Public participation

There are significant differences between Government and representative interests, the applicant for the licence and the community, in the extent they are allowed to influence, share or control the administrative decision-making process under the GT Act — particularly in relation to information disclosure. The Regulator's administrative decision-making process is not restricted to a consideration of advice received and written submissions. Other sources of information may be taken into account including (i) relevant independent research (GT Act, s. 51) or (ii) holding a public hearing in order to determine the licence application. However, the Regulator may give directions prohibiting or restricting the publication of evidence given, or material contained in documents produced at a public hearing (GT Act, s. 53). In terms of the community, a person other than the applicant for a licence has limited rights under the GT Act to access information held by the Federal agency at the time public submissions are sought. A request to the Regulator may only be made for a copy of the application and a copy of the prepared risk assessment and a risk management plan, subject to confidential commercial information being excluded: GT Act, s. 54.

Under the *Australian Risk Analysis Policy (2005)*, public comment is sought on the risk assessment and risk management plan prepared by the Regulator. Where deficiencies such as new risks, inaccurate assessments or better risk management strategies are identified in the consultation process, the risk assessment and risk management plan is updated and revised. Based on their analysis of the statutory and policy framework under the GT Act and Regulations, Lawson and Hindmarsh (2006) contend that the Regulator is not required to disclose any updated risk assessment and risk management plan that has taken into account any further advice or any written submission from the initial consultation process — notwithstanding that these materials may have been relied on by the Regulator in the administrative decision-making process.

(b) United Kingdom

The regulatory approach to biotechnology has progressed from control over the contained use of micro-organisms in laboratories (Directive 98/81/EC) and the *Genetically Modified Organisms (Contained Use) Regulations 2000* to control over the release of GMOs into the environment. Directive 2001/18/EC *on the deliberate release into the environment of genetically modified organisms* (the "Deliberate Release Directive"), repealed an earlier Directive (90/220/EEC). The Deliberate Release Directive was implemented into UK domestic law by the *Genetically Modified Organisms (Deliberate Release) Regulations 2002*. The Deliberate Release Directive sets out a notification procedure that must be adhered to prior to the release of any GMO. It also requires an environmental risk assessment to assess potential impacts on the environment and human health.

In their analysis of the EU Directives and complementary UK legislation, Thornton and Beckwith (2004) conclude that the approach to regulatory control of GMOs is based on the following principles:

- *The precautionary principle*: The preamble to the Deliberate Release Directive refers to the need for the precautionary principle to be taken into account in implementing the Directive. Moreover, the objective to protect human health and the environment when GMOs are deliberately released into the environment is to be in accordance with the precautionary principle (Article 1, Deliberate Release Directive). Additionally, there is a "*general obligation*" that all appropriate measures are taken to avoid adverse effects on human health and the environment which might arise when GMOs are released (Article 4).
- *Scientific risk assessment*: The preamble to the Deliberate Release Directive (para. 20) states that it is necessary to establish a common methodology to carry out the environmental risk assessment based on independent scientific advice. "*Environmental risk assessment*" [at Article 3(8)] is defined to mean "*the evaluation of risks to human health and the environment, whether direct or indirect, immediate or delayed*". The assessment is to be undertaken in accordance with the "Principles for the Environmental Risk Assessment" set out in Annex II of the Deliberate Release Directive. A general principle for environmental risk

assessment extends the analysis to "*cumulative long-term effects*". Cumulative long-term effects refer to: "*the accumulated effects of consents on human health and the environment including, inter alia, flora and fauna, soil fertility, soil degradation of organic material, the feed/food chain, animal health and resistance problems in relation to antibiotics*" *(Annex II)*.

- *The step-by-step principle*: The preamble to the Deliberate Release Directive (para. 24) states that the scale of release of GMOs is to be increased gradually, "*step by step*", but only if evaluation of the earlier steps in terms of protection of human health and the environment indicates that the next step can be taken.

Notification and consent requirements by regulatory agencies

As part of the standard authorisation procedure before a deliberate release of GMOs into the environment can be undertaken, a notification must be submitted to the national competent authority. The notification is to include (i) the technical information specified in Annex III of the Deliberate Release Directive as well as (ii) the environmental risk assessment and conclusions required in Annex II, together with (iii) any bibliographic reference and indications of the methods used (Article 6). If new information becomes available with regards to the risks of GMOs to human health and the environment, the information and conditions specified in the notification must be revised by the notifier [Article 13(6)]. An administrative decision by the competent authority, to give its consent, requires it to be satisfied that the release will be safe for human health and the environment.

Public participation: the Deliberate Release Directive

The public and, where appropriate, groups, are to be consulted on a proposed deliberate release (Article 9). A public inquiry may also be undertaken (Article 6). Member States are to make information on a proposed deliberate release of GMOs available to the public. Member States are required to set out arrangements for public consultation. In addition, the information contained in the notifications is subject to the confidentiality provisions of Article 25 (Article 9). A summary of the notification [Article 13(2)(h)], as well as assessment reports, are to be made available to the public indicating which GMOs should be placed on the market (including any conditions) as well as GMOs that should not be placed on the market: Article 14(3)(a). Subject to the qualification of confidentiality, the opinions of the Scientific Committee consulted are also to be made available to the public (Article 24).

The proposal for deliberate release is made available to the public before proceeding to decide the criteria and information requirements to be met for placing certain types of GMOs on the market, or in products. Public comments are sought and provided to the Committee set up for this purpose: Articles 16, 30. Where written consent has been given and new information becomes available which could have significant consequences for risk to human health and the environment, such information is to be evaluated by the competent authority and made available to the public (Article 8).

(c) The United States

No single, comprehensive Federal biotechnology statute has been implemented to provide a uniform process for evaluating risks or a uniform standard to facilitate administrative decision-making for risk management prior to commercialisation. Regulatory control of genetically engineered products arises under existing environmental, health and safety legislation. New regulations and policies have been developed for new biotechnology products, and introduced over time, to implement existing legislation.

Policy considerations and biotechnology regulatory control

The approach to regulatory control of GMOs has its origin in the Federal policy, the 1986 *Coordinated Framework for Regulation of Biotechnology* [the "*US Coordinated Biotechnology Policy (1986)*"]: 51 Federal Register 23302. A key feature of the *US Coordinated Biotechnology Policy (1986)* is the use of a risk-based system to ensure that the introduction of new biotechnology products is safe for human health and the environment (National Biological Information Infrastructure, 2007).

Under the *US Coordinated Biotechnology Policy (1986)*, Federal agencies responsible for the regulatory control of certain products became responsible for the same kinds of products that were developed through biotechnology (National Biological Information Infrastructure, 2007). The responsibilities of the Federal agencies are complementary and, in some instances, will overlap. Depending on the characteristics of the genetically engineered product, it may be subject to review by one or more of the agencies. If an organism is to be released into the environment, there may be three agencies which have oversight: the US Department of Agriculture ("USDA"), the Environmental Protection Agency ("EPA") and the Department of Human Services' Food and Drug Administration ("FDA"). The USDA is the lead regulatory agency for deliberate releases of genetically modified plants into the environment. The following summary of the responsibilities of Federal agencies for biotechnology is based on Turner (2001), Vogt and Parish (2001) and the National Biological Information Infrastructure (2007):

- The USDA's Animal and Plant Health Inspection Service ("APHIS") has responsibility for the planting of genetically engineered plants. Reviews for safety address whether the products are safe to grow;
- The EPA has responsibility over planting and food and feed uses of pesticides engineered into plants (or "plant-incorporated protectants"). In addition, it has responsibility for novel micro-organisms. Reviews for safety address whether the products are safe for the environment as well as whether a new use of a companion herbicide is safe;
- The FDA has responsibility over food and feed uses of all foods from genetically engineered plants. Reviews for safety address whether the products are safe to eat; and
- The US Department of the Interior will also have a significant role if a release into the environment has potential consequences for endangered species or migratory birds.

Relevant biotechnology legislation and environmental release of GMOs
The National Biological Information Infrastructure (2007) provides the following summary of the statutory framework for biotechnology:

i. The *Plant Protection Act* Title IV of the *Agriculture Protection Act. Regulations* issued by the US Department of Agriculture (7 C.F.R. Part 340) that deal with the introduction of genetically engineered products which are, or there is reason to believe they are, plant pests;
ii. The *Federal Insecticide, Fungicide and Rodenticide Act* ("FIFRA"), 7 USC Chapter 6. US Environmental Protection Agency Regulations including 40 C.F.R. Part 172 that deals with experimental use permits;
iii. The *Federal Food, Drug and Cosmetic Act*, 21 USC Chapter 9. The same US Environmental Protection Agency Regulations as for the FIFRA apply; and
iv. The *Toxic Substances Control Act* ("ToSCA"), 15 USC Chapter 53. Regulations issued by the EPA (40 C.F.R. 725) that deal with reporting requirements and review processes for micro-organisms.

Environmental release of GMOs: permits and review process
The importation, interstate movement or environmental release of certain genetically engineered organisms that could become pests must be authorised by APHIS through the issue of site-specific permits for field tests or for release into the environment. An application for a permit to release a genetically engineered plant into the environment must include the scientific information prescribed in 7 C.F.R., Section 340.4(b). Under the *US Coordinated Biotechnology Policy (1986)*, Federal agencies cannot impose further regulatory oversight, for example by requesting additional data,

unless this could be justified by the novel characteristics of the product (Levidow and Carr, 2000). However, the *US Coordinated Biotechnology Policy (1986)* provides two criteria where special evaluation is warranted: "intergeneric" (or new organism) and "pathogenic" (Turner, 2001). Intergeneric organisms are those formed by "*deliberate combinations of genetic material from sources in different genera*" whereas a "pathogen" is a virus or micro-organism having the ability to cause disease in other living organisms (51 Federal Register 23307).

The agency review process of a permit application is based on the NEPA process. The administrative decision may be to grant the permit and may include conditions to prevent the dissemination and establishment of plant pests; alternatively, the permit may be denied: 7 C.F.R., Section 340.4(e)(f). An environmental assessment is initially undertaken in order to evaluate the probable environmental impact of the release. Where the conclusion of the agency is a "Finding of No Significant Impact" ("FONSI"), a permit is issued (Vogt and Parish, 2001). An EIS will only be undertaken if the environmental assessment concludes that the release will have a "significant" impact on the environment and the planned action is a "major" action.

Alternative to permits: notification

The alternative to a permit application is a notification: 7 C.F.R., Section 340.3. Notification is an expedited procedure for approving limited permits that enable field-testing of a genetically engineered plant without a completed formal application that included an environmental assessment (Vogt and Parish, 2001). The notification procedure only applies where the genetically engineered plant meets all six prescribed eligibility criteria in 7 C.F.R., Section 340.3(b), e.g. the introduced genetic material is "*stably integrated* [as defined in 7 C.F.R., Section 340.1)]" in the plant genome. Also, all six prescribed performance standards in 7 C.F.R., Section 340.3(c) must be satisfied, e.g. no viable vector agent associated with the regulated article.

Following the completion of field-testing, an application is submitted. APHIS decides whether the product poses a risk of being a plant pest and whether it may be released on to the market. An environmental assessment is then undertaken by APHIS before the administration decision is made (Vogt and Parish, 2001).

Public participation

Jasanoff (1995) argues that the approach to biotechnology regulation in the United States has been one minimising the opportunities for the public to influence the administrative decision-making process. Support for this proposition reflects the reality that most field-testing involves the expedited procedure of notification, e.g. Vogt and Parish (2001) note that in 1998, "*99% of all applicants used the expedited process of notification*". There are no public comment requirements for notification (C.F.R., Section 340. 3) or permit (C.F.R., Section 340. 4) applications.

The essential position for public participation and biotechnology appears to be one that opportunities for maximising public inputs may only arise in specific circumstances; for example, when a permit is sought for the release into the environment of certain genetically engineered organisms and the environmental assessment process of the NEPA concludes that the release will have a "significant" impact on the environment and the planned action is "major". The EIA process would then follow. Whilst public participation is regarded as a significant cornerstone for the EIA process under the NEPA, the environmental assessment process does not have the same scope and opportunities for public participation compared with the EIA process.

(d) Preparation and Evaluation of the Risk Assessment

There are specific differences between each country in their approach to prepare and evaluate a risk assessment where there is to be an intentional release of a GMO into the environment. The primary distinction relates to the different groups assigned to do the risk assessment. In the UK, the applicant is required to submit a notification containing a complete environmental risk assessment for the proposed release (Article 13, Deliberate Release Directive). In Australia and the United States, the risk assessment is prepared by the Federal agency on the basis of information supplied by the applicant for a licence or permit. However, the function of evaluating the risk assessment is undertaken by independent advisory committees of scientists in the UK, by civil servants in the United States and opinion from groups of experts or individual experts or other regulatory agencies in Australia (EU–US Biotechnology Consultative Forum, 2000; Lawson and Hindmarsh, 2006).

BIOTECHNOLOGY: INTERNATIONAL ENVIRONMENTAL CONVENTIONS

International obligations on living GMOs have their foundations in the 1992 Convention on Biological Diversity. Agenda 21, the global future plan of action for the Rio Declaration on Environment and Development, is a significant foundation for implementing sustainable development, as it included principles for the "Conservation of Biological Diversity" (Chapter 15) and the "Environmentally Sound Management of

Biotechnology" (Chapter 16). Agenda 21 acknowledged that notwithstanding that biotechnology "*by itself cannot resolve all the fundamental problems of environment and development*" (para. 16.1), it had the potential to make a significant contribution by facilitating development in activities such as enhanced food security through sustainable agricultural practices, better health care and detoxification of hazardous wastes.

The parties to the Biodiversity Convention subsequently adopted a protocol on biosafety, the *2000 Cartagena Protocol on Biosafety to the Convention on Biological Diversity* ("the Biosafety Protocol"). The objective of the Biosafety Protocol (Article 1) is its specific focus on transboundary movements of genetically modified living organisms and their safe transfer, handling and use. This objective recognises that modern biotechnology may represent an adverse effect on the conservation and sustainable use of biological diversity as well as risks to human health. Furthermore, the objective of the Biosafety Protocol is required to be "*in accordance with the precautionary approach contained in Principle 15 of the Rio Declaration on Environment and Development*" (Article 1). Consequently, the Biosafety Protocol incorporates risk assessment and the precautionary principle in circumstances where administrative decisions are to be made by Government agencies whether or not to accept imports of genetically modified living organisms. Brunnée (2004) notes that the United States had sought to ensure that import restrictions of GMOs should be justified on risk assessments and scientific evidence, but not on precautionary grounds.

Legal Status of the Biosafety Protocol

The Biosafety Protocol establishes principles and procedures to facilitate decision-making by a signatory nation based on risk assessment and risk management. The Biosafety Protocol is only open to signature to Member States that are a party to the Biodiversity Convention. In terms of the Biosafety Protocol being legally binding on them, as at 1 May 2007, 141 nations had either "ratified" (80 nations) "approved" or "accepted" (5 nations) or "acceded to" (56 nations) the Protocol (Secretariat of the Convention on Biological Diversity, 2007b). Paradoxically, the United States, one of the world's largest producers of biotech food crops and animal feeds, is not a signatory to the Biosafety Protocol. The position for Australia, a nation that has ratified the Biodiversity Convention and implemented it into domestic law, is that it is a signatory, but is yet to ratify the Biosafety Protocol. In contrast, the United Kingdom ratified the Biosafety Protocol on 19 November 2003. By Regulation 1946/2003 (*on the transboundary movement of GMOs*) of the European Parliament and of the Council of 15 July 2003, the EU implemented the Biosafety Protocol. Existing EU legislation dealing with the import of GMOs, into and within the EU, is considered to be sufficient to comply with the Biosafety Protocol (Thornton and Beckwith, 2004). Accordingly, at this stage, legal obligations contained in the Biosafety Protocol have application in the UK only — and not in Australia or the United States.

LEGAL RIGHTS AS A SOURCE OF KNOWLEDGE POWER FOR NEGOTIATING SOLUTIONS

Mandel (2006) recognises two categories for possible legal challenges for biotechnology. First, a judicial review challenge for an administrative decision made by a Government agency to issue a permit or licence for the deliberate release of GMOs into the environment that may have the potential to cause, or have actually caused, adverse impacts on human health or the environment. In addition, a merits review involving civil enforcement actions where legislation may be breached resulting in potential or actual harm to human health or the environment. For example, where the release of a GMO into the environment may have the potential to harm a vulnerable or endangered species listed under environmental protection or threatened species legislation.

(a) Australia

Under the GT Act, there is a wide range of administrative decisions, in relation to the deliberate release of GMOs, that can be made by the Regulator and which may be subject to review. For example, an administrative decision to grant or to refuse a licence, with or without conditions, or to vary the licence may be subject to:

i. A judicial review under the AD(JR) Act before the Federal Court (GT Act, s. 183A); or
ii. A merits review before the Administrative Appeals Tribunal (GT Act s. 183). However, in marked contrast to a judicial review challenge, the opportunity for a merits review is restricted. The public are precluded from a merits review before the Administrative Appeals Tribunal. Statutory standing under the GT Act is expressly limited to an "*eligible person*" — a term defined to mean only the applicant for the licence or the licence holder (GT Act, s. 179).

Statutory standing under the AD(JR) Act is provided for an "*aggrieved person*" — a term that has a wide interpretation. The GT Act also extends standing to a State. A potential judicial challenge would primarily be based on whether there had been an error of law in the administrative decision-making process; or an unreasonable or improper exercise of power; or whether there was no evidence to support the findings of fact made by the administrative decision-maker.

In terms of the contention of Lawson and Hindmarsh (2006) that the Regulator is not required to disclose any updated risk assessment and risk management plan that may be relied on in the administrative decision-making process, an arguable case for a judicial review might be raised, in this situation. Procedural fairness may be in issue, given that the requirements for public input into the Regulator's decision-making processes under the *Australian Risk Analysis Policy (2005)* are "*clarity, transparency and accountability*". Based on established English common law principles for procedural fairness, an arguable case could be raised as to whether the consultation process had been "full and fair" to enable an "intelligent response" before the administrative decision to issue a licence was made. A proposition could be made that those who had been consulted initially had a legitimate expectation for being given a further opportunity to respond to any updated risk assessment and risk management plan, before the administrative decision was made. Legitimate expectation is a long-held accepted principle of the law of procedural fairness in Australia. Public law principles developed in the UK for legitimate expectation and information disclosure, as part of the public notice and comment period, would have persuasive value in Australia and so justify consideration when the environment is in issue.

There is also an alternative statutory pathway to the GT Act for the public to pursue legal rights arising from the deliberate release of GMOs into the environment. Specifically, there is the right to seek a statutory injunction under the EPBC Act before the Federal Court where there is a real chance that the release of the GMO may have the potential for a significant effect on a matter of national environmental significance; e.g. an endangered or vulnerable species and hence, biodiversity. The trigger for such a legal challenge is the meaning of the term "significant" — a term in which there is settled common law under the EPBC Act: *Booth v Bosworth* (2001) 114 FCR 39. Moreover, detailed policy guidelines exist for its meaning under the EPBC Act: *EPBC Policy Statement, Significant Impact Guidelines: 1.1 Matters of National Environmental Significance, May, 2006.*

(b) United Kingdom

As one of the general grounds of review, procedural impropriety is particularly relevant. The well-established common law principles in UK public law for full and fair information disclosure during the public notice and comment process are significant, given the adoption of the Aarhus Convention for public participation and the Deliberate Release Directive provide very broad public participation procedures that must be complied with as a legal obligation. Counter-balanced against this requirement is the fact that there is considerable scientific uncertainty inherent in the risk assessment methodology which may limit it as an objective, scientific method for evaluating risks to human health or the environment, e.g. Shere (1995). The dilemma for a Government agency is the need to comply with the public law principles for full and fair information disclosure to the public, in circumstances where the available information base may be at the limits of science, yet avoid the grounds for a judicial review challenge.

The principle of proportionality as a ground for judicial review, as well as risk assessment and risk management, were considered in *Pfizer Animal Health SA v Council of the European Union* [2002] E.C.R. II — 3305. The legal principles arising are relevant to risk regulation and GMOs. The case concerned a challenge related to the risk of transfer of resistance to antibiotics from animals to humans. The court concluded that the principle of proportionality required the measures adopted by institutions not to exceed the limits of what is necessary to obtain the legitimate objectives of the legislation in question. Where there is a choice between several appropriate measures, recourse must be had to the least onerous; in addition, the disadvantages caused must not be disproportionate to the aims pursued.

Scott (2004) has commented on the court's consideration of the principle of proportionality in the *Pfizer* case. In the context of risk regulation, proportionality encompasses a requirement to engage in cost-benefit balancing. Scott advances the proposition that "*even if public opinion is to be regarded as a factor which may be legitimately taken into account, it is critical also that it be allowed to resonate within the framework of a proportionality analysis*". Hilson (2005), in a review of EU case law, notes that there is no statutory power contained in the Treaty establishing the European Community, that results in administrative decision-makers being required to take public concern into account as a matter of legal obligation. Accordingly, it is Hilson's contention that it is only in the case where a public concern gives rise to a perception of risk that can be supported by scientific evidence, that public concern will be incorporated in the risk assessment process.

In an analysis of the decision in the *Pfizer* case, Scott (2004) drew attention to the following examples of the court's reasoning as to the procedural criteria for which regard will be had for assessing the legality of risk regulation:

- A scientific risk assessment undertaken must be based on the principles of excellence, transparency and independence.
- Although a decision-maker may disregard scientific advice, specific reasons for doing so must be given. The statement of reasons is required to be "*of a scientific level at least commensurate with that of the opinion in question*".

(c) United States

Litigation under the NEPA has been a catalyst for identifying critical issues of risk related to the release of GMOs into the environment. As Turner (2001) points out, litigation brought under the NEPA has been widely used to challenge (i) Federal Government agency decisions relating to biotechnology research and development, (ii) the *US Coordinated Biotechnology Policy (1986)* and (iii) administrative decisions approving activities for the release of GMOs into the environment. Grounds for a judicial review challenge for the administrative decision made under the permitting process for the deliberate release of GMOs into the environment have been founded on the grounds of procedural defects, the "arbitrary and capricious" test on the basis of substantial evidence as well as the "hard look" doctrine (Turner, 2001). As Turner points out, in most situations judicial review has taken place under the *Administrative Procedure Act* — notwithstanding that a NEPA challenge may be undertaken under other environmental statutes where the statute specifically provides for judicial review, e.g. the *Endangered Species Act*.

THE PRECAUTIONARY PRINCIPLE, SCIENTIFIC UNCERTAINTY AND ADMINISTRATIVE DECISION-MAKING

The origin of the precautionary principle is relatively recent and can be traced to a number of multilateral international treaties entered into by European countries in 1987 to protect northern hemisphere marine ecosystems from hazardous chemical inputs. In 1992, the precautionary principle was endorsed, on an international scale, through the signing by representatives of 172 nations of the *Rio Declaration on Environment and Development* (Rio de Janeiro, June 1992). The focus of Principle 15 of the Rio Declaration was on a precautionary approach to the protection of the environment: "*where there are threats of serious or irreversible environmental damage, lack of full scientific certainty should not be used as a reason for postponing cost-effective measures to prevent environmental degradation*".

The precautionary principle provides a guide to decision-making where scientific uncertainty exists. Excluding precaution may well mean that adverse environmental impacts would only be identified after they had occurred — when they may be irreversible (Lawrence and Taylor, 1990). In a very short period of time, the precautionary principle has increased enormously in scope and now extends to applications including sustainable development, environmental harm and the conservation of biodiversity and biotechnology; it has become incorporated into domestic environmental legislation and policies in many countries. In addition, the precautionary principle has appeared in almost all international instruments related to environmental protection, e.g. the Biosafety Protocol, and is considered to be the most developed form of prevention for use as a general basis for environmental protection (Kiss and Shelton, 2004).

(a) Australia

In May 1992, the Federal policy, the *Inter-Governmental Agreement on the Environment* (the "IGAE") was entered into by the Federal Government and all State and Territory governments. The IGAE incorporated the precautionary principle (at Clause 15.1), but extended the scope of Principle 15 of the Rio Declaration, by defining the manner in which the precautionary principle was to be applied through the use of risk assessment:

> "Where there are threats of serious irreversible environmental damage, lack of full scientific certainty should not be used as a reason for postponing measures to prevent environmental degradation.
>
> In the application of the precautionary principle, public and private decisions should be guided by:
>
> (i) careful evaluation to avoid, wherever practicable, serious or irreversible damage to the environment; and
> (ii) an assessment of the risk-weighted consequences of various options."

Following the introduction of this policy, the precautionary principle has been prescribed in Federal environmental legislation, e.g. the EPBC Act as well as a wide range of environmental harm-based legislation introduced by the States and the Territories. The GT Act prescribes the first element of the precautionary principle in the IGAE policy as part of the regulatory framework to achieve the statutory object [at s. 4(aa)]. It is particularly relevant that the object of the GT Act (at s. 3) is to protect the environment and health and safety of people "*by identifying risks posed by or as a result of gene technology and by managing those risks ... .*" These statutory provisions, together with the *Australian Risk Analysis Policy (2005)*, will require the application of the precautionary principle under the GT Act to depend on risk assessment and risk management being undertaken.

The application of the precautionary principle in environmental litigation by Australian courts has primarily been based on a literal interpretation of the principle. That is, by a consideration of uncertainty in the scientific evidence and the need to apply a cautious approach in order to avoid threat of serious or irreversible environmental damage in these circumstances. This approach taken in its first application [in *Leatch v National Parks and Wildlife Service* (1993) 81 LGERA 270] was also adopted in a more recent application in *BGP Properties P/L v Lake Macquarie County Council* [2004] 138 LGERA 237. In only one case, *Mohr v Great Barrier Reef Marine Park Authority & Others* (1998–99) 53 ALD 635, a merits review of a Federal agency administrative decision, was risk assessment considered to be an integral evidentiary requirement for precaution to be applied. In this case the Administrative Appeals Tribunal concluded:

> "We observe that an assessment of risk weighted consequences is necessary to ensure procedural fairness in the decision-making process where developmental interests (as reflected in this case by the mariculture proposal) and conservation interests, compete. Given the fact that no party has provided any evidence on the risk weighted consequences of the mariculture proposal, the Tribunal has not considered the operation of the precautionary principle to its decision making process."

(b) The United Kingdom

The Treaty establishing the European Community, a primary source of law in the EU, has specific provisions (at Title XIX) for environmental policy. These provisions and the precautionary principle were considered by the Court of First Instance in *Artegodan & Others v Commission of the European Communities* [2002] E.C.R. II-4945, which concluded:

> "As regards environmental matters, the precautionary principle is expressly enshrined in Article 174(2) EC ('the Treaty'), which establishes the binding nature of that principle. Furthermore, Article 174(1) includes protecting human health among the objectives of Community policy on the environment.
>
> Therefore, although the precautionary principle is mentioned in the Treaty only in connection with environmental policy, it is broader in scope. It is intended to be applied in order to ensure a high level of protection of health, consumer safety and the environment in all the Community's spheres of activity ... .
>
> It follows that the precautionary principle can be defined as a general principle of Community law requiring the competent authorities to take appropriate measures

> to prevent specific potential risks to public health, safety and the environment, by giving precedence to the requirements related to the protection of those interests over economic interests … "

Additionally, the Deliberate Release Directive makes it quite clear that its objective of protecting the environment and human health when GMOs are released into the environment is required to be in accordance with the precautionary principle (Article 1). A Communication from the European Commission "on the precautionary principle" (Brussels, 2 February 2000 — COM 2000 1) set out the manner in which the Commission intended to apply the precautionary principle when risks of environmental harm existed. Some of the features include:

- That the precautionary principle should be considered within a structured approach comprising the elements of (i) risk assessment, (ii) risk communication, and in particular (iii) risk management;
- That the implementation of an approach based on the precautionary principle should commence with a scientific evaluation, as complete as possible. That the degree of scientific uncertainty, at each stage, should be identified where possible;
- Recourse to the precautionary principle presupposes that a scientific evaluation of the risk makes it impossible to determine with sufficient certainty the risk in question — because of factors such as the insufficiency of the data or the imprecise, inconclusive nature of the data; and
- When the results of the scientific evaluation and/or risk assessment are available, all interested parties should be involved to the fullest extent possible in the study of various risk management options. The procedure should be as transparent as possible.

(c) The United States

Brunnée (2004) states that although the United States expressed concern over the scope of the precautionary principle in the Rio Declaration, it still accepted its inclusion in the Rio Instruments. However, from the time of the Rio conference, Brunnée contends that the United States has sought to contain the influence of the precautionary principle and has consistently denied its status as customary international law. Percival (2006) acknowledges that the regulatory approach of environmental law in the United States has generally been reactive rather than truly precautionary. However, Percival states that whilst the United States government has been reluctant to embrace the precautionary principle, far-reaching environmental legislation such as the NEPA, CWA and CAA and ToSCA is preventative in approach and endorses "*precautionary measures to prevent environmental damage before it occurs*".

(d) Precaution, Procedural Fairness and Administrative Decision-Making

Some of the criticism of the precautionary principle reflects its effect on the integrity of the administrative decision-making process. For the precautionary principle to be applied, it is sufficient to have reason to assume harmful environmental effects — rather than conclusive scientific evidence to provide a causal link (Lawrence and Taylor, 1990). In such a situation, the issue becomes whether the decision-making process is seen as being procedurally fair for parties having competing environmental or development interests. A literal interpretation of the principle based on a reason to assume environmental harm may not only be subjective, but also may result in administrative decisions being inordinately weighted to the environment. An effective risk analysis can offset this concern as it would facilitate procedural fairness in the administrative decision-making process when the precautionary principle is to be applied. Specifically, it would provide parties holding competing interests — development or the environment — with an equal opportunity to shape the final outcome (Christie, 1993d).

CONCEPTS IN RISK ASSESSMENT AND RISK MANAGEMENT: AN OVERVIEW

The statutory approaches for the regulatory control of GMOs in Australia, the UK and the United States all incorporate risk assessment and risk management as the basis for administrative decision-making. Power and McCarty (1997) acknowledge that science has played an important part in developing risk assessment techniques such as the selection of appropriate indicators for measuring and detecting potential change in the environment. However, they concede that science cannot address all the issues surrounding risk management, as policy must be used to make management-related decisions. Power and McCarty contend that whilst risk assessment is widely used, consensus is yet to emerge for an accepted comprehensive decision-making framework that clearly establishes the roles of science and policy in terms of formulating principles for environmental management.

(a) Risk

The release of GMOs into the environment is not without some risk. Risk is a multi-faceted term which creates problems in its interpretation. The

International Risk Governance Council (2005) recognises there are two dimensions to the interpretation of risk:

- A "factual" dimension which consists of outcomes which can be measured, e.g. the potential consequences and the likelihood of the occurrence of hazards that have been identified as a risk; and
- A "socio-cultural" dimension that reflects how a particular risk is viewed in terms of values and emotions. That is, whether society considers a risk to be acceptable, tolerable or intolerable is, in part, influenced by perceptions of impacts on societal values.

(b) Risk Assessment

Risk assessment involves an objective scientific evaluation of risks associated with the deliberate release of GMOs into the environment. Estimates of the likelihood of identified risks are combined with their consequences for human health or the environment. Risk is considered in the context of existing control measures.

(c) Quantitative or Qualitative Risk Assessment

Risk assessment quantifies the events and impacts that have the potential to occur. Mathematical techniques, such as predictive modelling, are used to determine the probability that a given hazard will occur and the consequences for environmental harm or for human health, if it were to occur. However, environmental risk assessment may also be qualitative because of factors such as the complexity of risk, the number of variables for which data are required and the level of detail required for administrative decision-making (*Australian Risk Analysis Policy, 2005*). Risk assessment can therefore be distinguished from an environmental impact assessment which, primarily, simply describes the potential environmental consequences of a proposed action.

(d) Risk Management

The objective of risk management is to reduce identified risks to an acceptable level of risk by evaluating various options for environmental management and protection as control measures to manage risk. Risk management facilitates administrative decision-making. It also involves a consideration of factors that may be difficult or impossible to quantify, e.g. social and policy considerations.

(e) Risk Analysis

Risk analysis is a broad concept that incorporates the processes of risk assessment, risk management and risk communication. Risk communication provides information about risk as the means for facilitating an understanding of risk. Additional objectives address the public perception of risk, reducing conflict and for achieving equitable conflict resolution outcomes (*Australian Risk Analysis Framework Policy, 2005*).

(f) Acceptability of Risk

Risk management recognises that all human activity involves some level of risk but that it is rarely possible to reduce risk to a zero level. The key outcome for risk analysis is deciding whether an identified risk can be managed, by applying control measures, to an acceptable level of risk. Whilst science has a clear and valid role to undertake a risk assessment, the issue that persists as a source of controversy is whether it is appropriate for science to determine the trans-scientific question for risk assessment of "how safe is safe enough?" (Christie, 1993d). The resolution of this issue involves judgements of risks against benefits as well as judgement about costs (Lawrence and Taylor, 1990). Furthermore, decisions by science as to what constitutes an acceptable level of risk may be significantly different from public opinion (Krimsky, 1993).

A common problem for the administrative decision-maker is where conflicts arise over risk between objective science and risk assessment and subjective public perceptions of risk. Such a situation represents the classic dilemma between the principles of expert scientific knowledge and democratic participation by the community in determining a standard for an acceptable level of risk (Krimsky, 1993). Deriving a standard for an acceptable level of risk for human health, public safety or the environment continues to be a vexing issue, not only for administrative decision-making, but also for public trust in the Government agency responsible for the regulatory control of risk.

REGULATORY APPROACHES TO RISK ASSESSMENT

The United States National Academy of Sciences (1983) has been the key source for guidelines on the framework for undertaking a health risk assessment. Health risk assessment proceeds along the following four stages: (i) hazard identification, (ii) dose–response assessment, (iii) exposure assessment and (iv) characterisation of risk. The National Academy of

Sciences saw the progressive development of the science of risk assessment as a means for offsetting any limitations in risk assessment, rather than to replace it. The statutory procedures for risk assessment in Australia, the UK and the United States reflect this objective.

(a) Australia

The standard approach for risk assessment undertaken by the Federal agency involves the following elements and in the following sequence of steps (*Australian Risk Analysis Policy, 2005*):

i. Hazard identification;
ii. Identification of a pathway or route of exposure. A causal link between the hazard and an adverse outcome for human health or the environment needs to be demonstrated;
iii. The consequences, magnitude and effects on human health, or the environment, for a hazard identified as a risk. This may include a dose–response assessment;
iv. The likelihood of the occurrence of each identified hazard — expressed as frequency and probability;
v. Estimation, or characterisation of the risk. Scientific evidence is used to estimate the level of risk based on a combination of the likelihood of risk and its consequences for potential harm. The risk estimate is used to consider the control measures that may be used to manage those risks;
vi. Risk management. Identification of risks that require management as well as control measures that could be used to effectively manage the risks to an acceptable level in such a way to protect the health and safety of people and the environment; and
vii. Risk communication.

(b) The United Kingdom

Annex II of the Deliberate Release Directive described the principles for environmental risk assessment. The pre-condition for risk assessment is the characterisation of GMOs and releases. Consideration is given to the intended release, or use, in terms of factors such as the scale, the potential receiving environment and the combination of these. The prescribed steps in the risk assessment methodology are summarised as follows:

i. Identification of the characteristics of GMOs which may cause adverse effects on human health or the environment — recognising that potential adverse effects of GMOs will vary from case to case;

ii. Evaluation of the potential consequences of each adverse effect, if it occurs, in terms of the magnitude of the consequences;
iii. Evaluation of the likelihood (or probability) of the occurrence of each identified potential adverse effect. Consideration is given to the characteristics of the environment and the manner of release;
iv. Estimation of the risk posed by each identified characteristic of the GMO(s). Scientific data on the likelihood of the adverse effect occurring are combined with the magnitude of the consequences, if it occurs;
v. Application of strategies or control measures for managing risks arising from the deliberate release of GMOs. Where risks are identified that require management, a risk management strategy should be defined; and
vi Determination of the overall risk of the GMOs, taking into account any proposed risk management strategies.

(c) The United States

From the outset, the *US Coordinated Biotechnology Policy (1986)* identified a number of categories required for risk assessment: information on exposure, environmental fate and effects on human health and the environment. Turner (2001) states that when an environmental assessment for a regulated article is undertaken under the NEPA, the approach taken by APHIS to determine the outcome of an application for a permit for field tests or release into the environment, is to ensure consistency with the biosafety procedures in Annex 2 of the *"United Nations Environment Program Guidelines for Safety in Biotechnology"*. That is, by addressing steps such as:

i. The identification of hazards;
ii. Assessment of actual risks that may arise from the hazard;
iii. Determining how the risk can be managed and whether to proceed with the proposed action; and
iv. Comparing the risk from the proposed action with risk from actions involving comparable organisms.

SCIENTIFIC UNCERTAINTY AND RISK ASSESSMENT

The accuracy and validity of risk assessment, as is the case with environmental impact assessment, is a continuing source of scientific debate. This concern recognises that complex interrelationships are frequently the subject of risk assessment, which will be problematic should the available scientific data be limited or inadequate. To deal with uncertainties, the science of risk assessment makes use of statistics and probability theory;

probability is the mathematical measure of risk. In addition, mathematical techniques such as predictive modelling are used to quantify the probability of risk (Brown, 1992).

(a) Sources of Scientific Uncertainty

McClenaghan *et al.* (2002), Thornton and Beckwith (2004) and Lawson and Hindmarsh (2006) all challenge risk assessment as an objective, scientific method for evaluating risks to human health or the environment where the following limitations, inherent in the methodology, are present:

i. A basic insufficiency of objective scientific data. The limitations in scientific knowledge results in uncertainty in the risk assessment;
ii. Lack of methodologies for key steps in the process, e.g. dose–response assessment and exposure assessment;
iii. Problems in ensuring consistency and equal levels of professionalism and expertise across highly complex areas;
iv. An inherent uncertainty in predicting potential future risk;
v. A failure to acknowledge subjectivity through the use of value judgements to identify hazards and to assess risks; and
vi. Uncertainty extends to the public perception of risk making consensus as to what constitutes an objective, acceptable risk problematic.

Moreover, McClenaghan *et al.* (2002) contend that the limitations in risk assessment methodology also extend to risk management as this process is "*equally subject to the value judgements and guesswork that are central to the 'science' of risk assessment*".

(b) Scientific Uncertainty: Risk Assessment and GMOs

Ramirez and Galindo (2004) highlight the experimental problems for risk assessment and GMOs: the number of variables associated with a GMO and its derivates is so high that there are not enough scientifically designed studies to test the extent and impact of each one of them. Consequently, in these circumstances, there will be uncertainty in the risk assessment. For example, the most basic issue persists: whether all hazards — an event, organism or substance that may become a risk to human health or the environment — have been identified. In addition, the deliberate release of GMOs into the environment may be associated with unknown, novel risks and/or predictable harm relative to their normal, i.e. non-GMO, counterparts (Levidow and Carr, 2000).

Because the sheer volume of research needed to test the extent and impact of all possible hazards is prohibitive, value judgements lead to shortlisting and selection of events to be included in the risk assessment. In this regard, the scientific basis for any assumptions that lead to the shortlisting is critical. Any subjectivity introduced, by relying on value judgements, needs to be acknowledged.

(c) Scientific Uncertainty: Mathematical Modelling and GMOs

Ramirez and Galindo (2004) refer to the use of novel dynamic mathematical models to address the complex biological processes associated with GMOs and to quantify the probability of risk. However, mathematical models may be of varying accuracy and so add a further dimension to the uncertainty of risk assessment. The sources of uncertainty in the reliability of the output from a mathematical model relate to the extent that accepted procedural steps for constructing a mathematical model have been adhered to. Key procedural steps include:

i. The experimental database and any underlying assumptions relied on to shortlist and select events to be included in the risk assessment model;
ii. An independent validation of the accuracy of the model output carried out against field data other than the data upon which the model was constructed; and
iii. Some form of sensitivity analysis to address any uncertainty in model output associated with the equations used in the model that may arise from sources such as lack of data or of an understanding of key driving processes and variables.

The common law principle in the Australian case, *Lawrence v Kempsey Shire Council* (1995) 87 LGERA 49, reflects one essential scientific standard for model development:

> "As to the scientist's computer model of how Christmas Creek would behave under certain conditions; unless there is some factual evidence or satisfactory expert opinion evidence to support the assumptions fed into the computer, the result obtained from the computer has no evidentiary value."

PUBLIC PERCEPTIONS AND RISK ASSESSMENT

The meaning of risk incorporates a factual dimension and a socio-cultural dimension (International Risk Governance Council, 2005). In relation to the latter, Perhac (1998) contends that as value judgements are required throughout the risk assessment process, assessment of risk should embrace the "*widely maintained*" view that the public are an appropriate source of the judgements to be made in risk assessment. Furthermore, Perhac considers that knowledge, in risk assessment, is not the sole province of science, the regulatory agency or risk assessors; relevant, possibly unique, knowledge also resides with the public. Support for this proposition also emerges from the review by Fiorino (1990), specifically, that non-experts identify problems, issues and solutions that experts may miss (Isaacson, 1986). An example given is that experts relying solely on diagnostic and preventative methods would have missed risks to human health from Agent Orange and diethyl stilbestrol ("DES"). In addition, studies of lay judgements about technological hazards indicate sensitivity to social and political values that the models of experts would not acknowledge. The lay public may, in fact, have better capacity than experts alone (Barber, 1984).

The analysis by Barnes (2001) is important as it clearly illustrates the essential linkage between public perceptions of risk, public participation and administrative decision-making by a Government agency in relation to environmental risk: the public have concerns about the safety of the wider environment and their near surroundings; the regulatory agency has a need helping to provide a safe environment. A failure to counter-balance these interests in the regulatory agency's decision-making process will erode its credibility. A decline in public trust and disbelief in the agency would follow as a consequence.

However, as Barnes points out, the problem for administrative decision-making for risk is not only one of failing to understand the belief systems that underlie the public perception of risk, but also one that reflects a failure by the agency to engage the public effectively.

PARTICIPATORY DECISION-MAKING AND SHARING POWER: RISK AND DELIBERATE RELEASE OF GMOS

The International Risk Governance Council (2005) states that existing participatory approaches often fail to effectively integrate and involve all stakeholders and the general public within the processes by which information is collected, analysed and communicated and upon which administrative decisions are made. This proposition is considered in the context of the participatory approaches in Australia, the UK and the United States.

(a) Existing Participatory Approaches

The models for public participation and the regulatory control of GMOs reviewed for Australia and the United States reveal that the focus of public participation is primarily on notification and public comment procedures. It should be emphasised that the legislative procedures in the United Kingdom are far broader than the more limited procedures in Australia and the United States. The Aarhus Convention for public participation, together with the Deliberate Release Directive, provides the UK with greater flexibility in public participation procedures for notification, public comment and, most importantly, information disclosure. Whilst there is some degree of shared involvement between the regulatory agency and the public in the notification and public comment procedures in all three jurisdictions, participatory power does not extend to joint analysis and control over decisions and their implementation in the administrative decision-making process for the regulatory control of biotechnology.

This limitation in public participation for biotechnology and risk assessment also needs to be considered in the context of a recommendation made by the EU–US Biotechnology Consultative Forum (2000). That is, an inclusive regulatory system would enable administrative decisions to be made in a manner that respects community judgements of appropriate growth goals, value judgements and ethical boundaries. Or, as Fiorino (1990) states, the participatory approach to risk assessment should enable the public to participate more effectively, and on a basis of greater equality with the regulatory agency and scientific experts.

Consequently, a consideration of some of the alternatives to the existing participatory approaches is warranted in terms of the extent that they may offset these limitations and provide the community with an adequate basis of power to participate effectively in the administrative decision-making process over risk to the environment or public health.

(b) Alternative Participatory Approaches

Fiorino (1990) reviewed five participatory approaches in terms of the extent the public are able to share in collective decision-making for environmental risk: citizen review panels; initiatives (where citizens place issues on the ballot for voter approval); negotiated rule-making under the Negotiated Rule-Making Act (an ADR process); public hearings; and public surveys. The conclusions from this study are summarised as follows:

i. The best example of shared authority in environmental risk decision-making was negotiated rule-making, where all of the parties are required to reach agreement by consensus;
ii. Review panels, public surveys and public hearings offered little or no opportunity to share authority;

iii. Negotiated rule-making and review panels, relative to the other approaches, had significant strengths for engaging people in collaborative problem-solving, for providing opportunities for sharing of information, in dispute analysis and for better understanding each others' needs and concerns;
iv. The distinction between negotiated rule-making and review panels, in terms of having power to share authority in decision-making, was a significant difference. Concern was expressed that a panel review may raise expectations of possible influence that cannot be met;
v. Although initiatives enabled citizens to exercise full decision authority, a single person has limited influence over the outcome of the ballot; and
vi. A majority vote in an initiative forces a decision from sharply divided choices, rather than through consensual agreement.

(c) ADR and Interest-Based Negotiation

The conclusions of the Fiorino (1990) study clearly illustrate the significant advantages of a structured ADR process, in this case negotiated rule-making, for the assessment of environmental risk compared with the other approaches. All of the features of negotiated rule-making identified in this study — sharing of power in decision-making and collaborative problem-solving, sharing of information, dispute analysis, understanding interests of all parties and agreement by consensus — are also present in many other ADR processes such as facilitative and evaluative mediation. However, the use of ADR as a participatory approach is not mandated in the participatory approaches for biotechnology and risk assessment in Australia, the UK or the United States.

A participatory approach based on ADR and interest-based negotiation, in which the regulatory agency is prepared to share its power with representative interests, incorporates and extends the above advantages. Interest-based negotiation enables the regulatory agency to better understand the belief systems that underlie the public perceptions of risk associated with the deliberate release of GMOs into the environment, i.e. the socio-cultural dimension of risk. Consequently, a participatory approach based on ADR and interest-based negotiation would facilitate achieving the goal of involving the public in a meaningful way in the assessment and management of biotechnology and risk and so offset any perceptions of manipulation by the Government agency in the participatory process.

FINDING SOLUTIONS FOR MANAGING AND RESOLVING INFORMATION CONFLICTS FOR BIOTECHNOLOGY AND RISK

The International Risk Governance Council (2005) has identified risk-related decision-making as often failing to successfully link scientific knowledge with the socio-cultural issues associated with risk. As a result, this leads not only to eroded public confidence and trust in the Government agency, but also diminished returns on, or loss of, investment. Accordingly, the International Risk Governance Council contends that a dual approach is needed for risk management which provides "*equal room for both scientific evidence and value-based perceptions of risk*". This need for a dual approach is considered in the context of using ADR and interest-based negotiation as the participatory approach for managing and resolving information conflicts associated with risk assessment and management and biotechnology.

(a) Risk Assessment and Conflict Management

Following the public comment period, scoping would be undertaken by the regulatory agency to identify and prioritise the issues in dispute. Hazard identification would then be undertaken by representative parties, with the goal of reaching agreement by consensus on the hazards having the potential to cause harm to human health or to the environment.

Next, risk assessment would be undertaken using the scientific round table. Data mediation would be used to identify the available scientific information that could be used for the risk assessment and to ensure that all key pieces of information had been disclosed. Areas of scientific uncertainty and their implications for risk assessment would need to be identified. Consideration would then be given to the form of risk assessment to be undertaken — quantitative or qualitative.

Given that empirical studies have documented significant differences between expert and lay perceptions of risk (Douglas, 1985), the scientific round table could prepare two risk scenarios that would be subject to risk assessment — one scenario where hazards identified as a risk were based on scientific evidence, the other based on hazards identified as a risk by the public (Perhac, 1998). Different perspectives on risk associated with the same application for a licence or permit for the deliberate release of GMOs, in terms of the effects on human health or the environment, would ensure scientific and public perceptions of risk are given equal and effective consideration. The significance of the scientific round table is that the assessment of the risk scenarios prepared would involve all the scientific experts for the representative parties in a joint problem-solving exercise, having access to the same information and agreeing to the underlying assumptions that would be used to estimate or characterise risk in each scenario.

The scientific round table would also be required to reach agreement by consensus on objective criteria that could be used to rank risks in order to identify the risks requiring management, e.g. by consideration of adverse impacts, over the short and long term; in relation to criteria such as severity (or intensity), extent (area), cumulative effects, biodiversity, sustainability and resilience. The scientific round table would then evaluate the options, or control measures, for the risks identified as requiring management. Objective criteria for evaluating the control measures should be the subject of agreement by consensus at the round table, e.g. effectiveness, economic and technical feasibility, policy guidelines and avoidance of adverse impacts on ecological or human health. The outcomes from the scientific round table should be provided to the representative parties, for consideration, at the next phase — conflict resolution.

(b) Risk Management and Conflict Resolution

The objective of conflict resolution is to find a solution, by consensual agreement, for managing the risks to human health and the environment posed by a proposal for the deliberate release of GMOs into the environment, at an acceptable level of risk. Possible conditions to attach to the risk management plan may be arrived at by agreement by consensus. This phase involves the representative parties and the use of the ADR process of evaluative mediation and interest-based negotiation.

The representative parties would need to reach consensual agreement on objective criteria which could be used to decide whether a risk is acceptable. One approach is that of Brown (1992) in which the standards for evaluating and comparing risk are based on different "risk equations" and include:

- *Elevated (or added) risk*: Refers to an increase in risk resulting from the hazard. A comparison is made of the identified risk with the additional hazard arising over the natural background;
- *Balanced risk*: A comparison of the identified risk with practicable, applicable alternatives, e.g. management options or control measures to mitigate risk. A weighting of each risk versus benefit may be undertaken for each alternative;
- *Benefit risk*: A comparison of the identified risk with the benefit it may bring. Probably applicable only to a risk judged to be tolerable, i.e. a risk from activities that are prepared to be accepted in order to obtain benefits.

- *Comparative risk (based on a common denominator, usually death)*: A comparison of the risk with the probability of death published in actuarial life loss tables.

In terms of biodiversity risk, an additional standard recognised by science is the "*quasiextinction risk*" (Ginzburg *et al.*, 1982). The basis for quasiextinction risk, for a species, is that there is a chance that a small population threshold may be crossed to a lower level that may be unacceptable for either conservation, management, economic or aesthetic purposes. It is a flexible term as it refers to all kinds of decline rather than only to a decline to some critical level (Burgman *et al.*,1993).

(c) Precaution, Uncertainty and Permit Conditions

In the context of biotechnology, the application of some form of bond may warrant consideration as a condition imposed on a permit or licence approving an application for the deliberate release of GMOs into the environment. The performance bond has had wide application in administrative decision-making by Government agencies and courts as a condition for approving projects or activities that may have the potential to cause environmental damage.

Costanza and Perrings (1990) have advanced the concept of an incentive-based model to manage the environment for precaution under uncertainty — the "*flexible environmental assurance bonding system*". It is a variation of the performance bond and is designed to incorporate both known and uncertain environmental costs into the incentive system and to induce positive environmental technological innovation. An assurance bond is levied equal to the current best estimate of the largest potential future environmental damages arising from a proposed action seeking legislative approval from a Government agency in the form of a licence or permit. The assurance bond would be kept in an interest-bearing account for a predetermined length of time.

In keeping with the precautionary principle, the approach requires a commitment of resources, up front, to offset the potentially catastrophic future effects of current activity. Portions of the bond (plus interest) would be returned if, and when, the holder of the permit could demonstrate that the suspected worst case damages had not occurred, or would be less than was originally assessed. Where environmental damages had occurred, portions of the bond would be used to rehabilitate or repair the environment; in addition, to possibly compensate injured parties.

LITIGATION V NEGOTIATION AND BIOTECHNOLOGY CONFLICTS

A conflict over biotechnology and risk will involve the same elements: questions of law, facts, policy and values. However, the status and role of each of these elements in arriving at a litigated or negotiated outcome, as well as the process for resolution, can be distinguished when risk is a central issue.

There are a number of significant limitations for the legal process and adjudication when biotechnology and risk are in issue. The risk assessment process involves the evaluation of numerous and complex facts, including many assumptions and unknowns associated with hazards identified as a risk and their environmental consequences. Scientific uncertainty in the factual evidence will almost certainly prevail. In this situation, litigation will resolve divergent expert opinion evidence in circumstances where uncertainty, or even ignorance, exists as the final determination of facts is crucial for the legal process (Martin, 1977).

Moreover, risk management and the concept of risk acceptance — both pivotal concepts for administrative decision-making — incorporate objective science as well as a socio-cultural dimension that reflects values and public perceptions of risk. Not only are values and perceptions subjective, but they are also not matters that a court must take into account as a matter of legal obligation where they have not been expressly or impliedly provided for in a statute. Furthermore, courts have less flexibility, compared with ADR, for balancing the broad public policy considerations that affect competing interests when decisions have to be made over the management of risks associated with biotechnology.

The advantage of ADR and interest-based negotiation, in circumstances of scientific uncertainty and where public perceptions of risk need to be understood, is that it involves a process of shared responsibility through joint fact-finding, joint problem-solving and agreement by consensus — all features that are not part of litigation. ADR and interest-based negotiation provides the dual approach for risk management advocated by the International Risk Governance Council (2005) as it provides "*equal room for both scientific evidence and value-based perceptions of risk*". Interest-based negotiation enables values and perceptions of risk to be effectively understood by all parties. In addition, ADR, unlike the legal process, integrates the analysis of scientific knowledge with a consideration of value-based public perceptions and concerns of risk, e.g. whether they were speculative or real. This linkage is the essential foundation of administrative decision-making for risk management.

Additionally, ADR and the interest-based approach to problem-solving provides the regulatory agency, the scientific community, the corporate sector, the risk innovator and the public with an adequate basis of power for all parties to effectively participate. Finally, where a regulatory agency shares its power in the administrative decision-making process, it follows that solutions sought as part of the process of negotiation need to, at least, consider the obligations imposed by biotechnology legislation, where the interpretation of the law is not in dispute. Consideration of all of these characteristics leads to one inevitable conclusion: ADR is a more effective process for managing and resolving conflicts over biotechnology and risk, compared with the legal process and the courts.

8. Hazardous chemicals and public health

> "All substances are poisonous … the right dose differentiates a poison from a remedy: Paracelsus 1493–1551."
>
> Quoted in **Klaarsem *et al.* (1986)**

> The Swiss chemist, Dr Paul Hermann Müller (1899–1965), won the Nobel Prize in Physiology or Medicine in 1948 "for his discovery of the high efficiency of DDT as a contact poison against several arthropods."
>
> **The Nobel Foundation (2006)**

> "It is not my contention that chemical insecticides must never be used … I contend that we have allowed [poisonous and biologically potent] chemicals to be used with little or no advance investigations of their effect on soil, water, wildlife and man himself. Future generations are unlikely to condone our lack of prudent concern for the natural integrity of the natural world that supports all life."
>
> **Rachel Carson (1962)**

> "The current regime [for disposal of toxic wastes is one] of impasse, conflict, fragmentation and futile attempts at coercion."
>
> **Gerrard (1994)**

INTRODUCTION

A heightened awareness of the problems associated with chemicals and the environment emerged with the publication of *Silent Spring* by Rachel Carson in 1962. As Moore (1987) points out, Rachel Carson did not advocate the cessation of chemical pesticide use. Rather, attention was drawn to the fact that a large number of "*poisonous and biologically potent chemicals* [are being put] *indiscriminately into the hands of persons largely or wholly ignorant of their potential for harm*" and were possibly making the earth unfit for life. At that time, only about 500 chemicals were introduced each year. A review of the extent of chemical use by Watfa *et al.* (1998) clearly illustrates the significant change in the global situation: about 70,000 chemical substances are produced for regular use in the industrial sectors, agriculture and for other needs. However, only a few thousand of these chemical substances are produced in substantial quantities for industrial use; of the chemicals which are marketed, 5 to 10 per cent are considered hazardous; 150–200 chemicals are known carcinogens; approximately 2000 chemical substances are on the "*List of Dangerous Substances*" in the European Union; in the United States, 2000–3000 substances are notified each year under the *Toxic Substances Control Act* of 1976. In addition, on a global basis, around 80 per cent of chemicals are consumed by the developed world. The remaining 20 per cent is consumed by developing countries. However, the trend in developing countries is one of an increasing demand for chemicals — with predictions that the future demand for chemicals will increase more rapidly in developing countries than in developed regions (Watfa *et al.*, 1998; Buccini, 2004).

When considered in the context of improving quality of life and standard of living, the technological revolution associated with the production and use of chemicals has provided significant benefits for society. Nevertheless, there are costs arising from the improper use and disposal of the large number of chemicals in use and public health problems. Toxic chemicals include carcinogens (cancer-producing substances), teratogens (substances affecting the embryo) mutagens (substances causing genetic changes), irritants and systemic poisons through to chemicals with dermatological effects (Watfa *et al.*, 1998). From a public health perspective, Tynan (2005) states that persistent organic pollutants are of particular concern because they may be transported great distances from their source, degrade very slowly and bioaccumulate in humans. The World Health Organisation (2007) has clearly identified some of the problems associated with the release of toxic chemicals and their impacts on public health. For example, acute exposure to toxic chemicals can lead to death or serious illness. Similarly, chronic exposure to toxic chemicals over the long term can increase the risk of developmental and reproductive disorders, immune-system disruption, endocrine disruption, impaired nervous-system function and the development of certain cancers. Of particular concern is that exposure of children to toxic chemicals places them at a higher risk of serious illness compared with adults.

It is clear that the problems created by exposure to hazardous chemicals in the environment represent major potential risks to public health and may arise under a wide range of potential exposure situations (Lippmann and Thurston, 1988; Buccini, 2004), including:

i. Hazardous waste generation, transportation, disposal and storage (including landfills);

ii. Environmental contamination (e.g. waters and the atmosphere) following the escape or spillages of chemicals; and
iii. The production of unintentional by-products, e.g. "chemical cocktails" where chemicals intermingle in a hazardous waste dump.

There are two alternative scenarios for a hazardous chemical exposure–public health conflict. The first is one in which it is alleged that exposure to a specific chemical has actually caused a public health problem. Such a scenario frequently arises in an occupational health setting. It is the basis for civil litigation in either a common law negligence action or a personal injury claim under workers' compensation legislation. Litigation, in this case, is a reaction to a public health problem that has already occurred. The alternative scenario is one in which it is claimed that a specific chemical may have the potential to cause a public health problem, e.g. a proposal for a hazardous waste disposal at a landfill or the construction of a hazardous waste facility. However, an actual health problem has not yet occurred. In this situation the focus is on the resolution of the conflict in order to avoid or offset a potential public health problem.

As Watfa *et al.* (1998) point out, chemicals are here to stay and so we must learn to live with them safely. This statement needs to be considered in the context that 7 per cent of global income and 9 per cent of international trade arises from the global chemicals industry (Buccini, 2004). The reality of the situation for global chemical use therefore highlights the importance of effectively applying some of the elements of the "*obligation to endure*", as identified by Rachel Carson almost half a century ago and which have stood the test of time: i.e. an increased awareness of the nature of any threat from the use of chemicals and, as the community is being asked to assume the risks, the community having a role in the decision-making process with all information and facts in their possession.

GLOBAL CHEMICAL USE AND SUSTAINABLE DEVELOPMENT

Buccini (2004) considers that the key contemporary problem confronting the global use of chemicals is, in fact, a "*classic sustainable development issue*". Quite clearly, this proposition is correct as the issue is one of a question of balance, i.e. whether modern society can counter-balance the multiple objectives of:

i. Deriving benefits from industrial growth and the associated use of chemicals; with

ii. Offsetting the immediate and long-term risks to public health and the environment imposed by the use and disposal of chemicals.

Whilst this chapter has as its focus, chemicals and public health, it should also be evident that environmental contamination associated with chemical use may also impact on wildlife (Buccini, 2004). Chemical pollution has been identified as a contributing factor to the decline in global biodiversity (IUCN, 2006). Therefore, problems related to chemical use and sustainability are equally as relevant to ecological health as they are to human health. "Ecological health" has been defined in terms of four major elements — *sustainability,* which is a function of *activity*, *organisation* and *resilience*, as: "*An ecological system is health free from 'distress syndrome' if it is stable and sustainable — that is, if it is active and maintains its organization over time and is resilient to stress*" (Haskell *et al.*, 1992).

LEGISLATIVE APPROACHES FOR THE REGULATORY CONTROL OF HAZARDOUS CHEMICALS

The traditional approach for dealing with environmental problems associated with chemical substances has been a reactive one aimed at controlling the impacts of pollution through legislative control. However, a contemporary global approach has seen a shift away from control of pollution to a problem-solving approach directed at preventing harm to the environment. The concept of environmental harm incorporates pollution but extends to the much broader considerations of environmental quality (Bell and McGillivray, 2006). Environmental harm may arise when chemical substances come into contact with any medium — air, soil, vegetation or water — resulting in actual or potential harm to human health or ecological health. Bell and McGillivray (2006) distinguish between the "*First Generation*" approach of control, with its focus on polluting industries and control, and the "*Second Generation*" approach. The second generation approach is directed at a shared responsibility between Government, industry and the community to address environmental harm.

Any proposed use of a site for an industry, activity or process that may have the potential to cause environmental harm requires an administrative decision by a Government agency exercising statutory power, to decide whether or not to grant approval. The form of statutory approval is variously referred to as a licence, permit, environmental authority or authorisation, environmental protection licence or a best practice licence.

(a) Australia

The only Federal legislation dealing with hazardous waste is the *Hazardous Waste (Regulation of Exports and Imports) Act 1989 (Cth)* that regulates the export, import and transit of hazardous waste. Its object is to ensure that hazardous waste is managed in an environmentally sound manner to protect humans and the environment, both within and outside Australia, from the harmful effects of the waste. Regulatory control of the treatment, storage, disposal and management of hazardous wastes in Australia arises under environmental protection legislation enacted by the States and the Territories — and not Federal legislation.

Environmental harm based regulatory control

Environmental harm, as the second generation approach for regulatory control of the environment, was first introduced in South Australia in 1993: the *Environment Protection Act 1993 (SA)* ["EP Act (SA)"]. Almost all of the other States and both Territories then followed: *Environmental Protection Act 1994 (Qld)* ["EP Act (Q)"]; *Environmental Management and Pollution Control Act 1994 (Tas)* ["EMPC Act (Tas)"]; *Protection of the Environment Operations Act 1997 (NSW)* ["PEO Act (NSW)"]; *Environment Protection Act 1997 (ACT)* ["EP Act (ACT)"]; and *Waste Management and Pollution Control Act 1998 (NT)* ["WMPC Act (NT)"]. However, in Victoria, the *Environment Protection Act 1970 (Vic)*, whilst still primarily pollution-based, has provisions giving general effect to "harm to the environment". In Western Australia, the *Environmental Protection Act 1986 (WA)* is also pollution-based but does now also incorporate "environmental harm" [s. 3A(2)].

For the States and Territories that have adopted an environmental harm-based approach to regulatory control, the statutory object of the legislation for environmental protection is complemented with the need to maintain, or to allow, ESD: EP Act (SA), s. 10; EP Act (Q), s. 3; PEO Act (NSW), s. 3; EP Act (ACT), s. 2; and the WMPC Act (NT), s. 5. In Tasmania, the legislation incorporates the objectives of, (i) sustainable development for resource management and planning and (ii) ecosystem sustainability for environmental management and pollution control [EMPC Act (Tas), s. 109, Schedule 1]. The pollution-based legislation of Victoria now incorporates the principle of ESD as a statutory purpose [EP Act (Vic), s. 1B], but this not the case for the *Environmental Protection Act* in Western Australia.

Characterizing hazardous wastes: environmental harm

The focus of harm-based regulatory control is directed at the potential or actual impacts of hazardous waste to cause environmental harm.

Consequently, hazardous wastes are characterised in terms of their likelihood to cause environmental harm if, for instance, they are improperly stored, disposed of or managed. This approach to characterising hazardous waste differs from the UK and the United States. Environmental harm, as defined in legislation, refers to actual or potential adverse effects on environmental quality as well as on human health or safety. Additionally, harm is sometimes defined in terms of the costs of environmental damage that are incurred, e.g. following the escape of hazardous waste, and the cost of damage exceeds threshold amounts prescribed in the legislation.

Approvals for waste management

The environmental protection legislation for each State and Territory prescribes those industries, processes and activities that require statutory approval. Examples for the harm-based jurisdictions include chemical plants and the transportation and disposal of hazardous wastes. Regulatory control of environmental harm associated with waste disposal and waste recycling is achieved by imposing a requirement, depending on the jurisdiction, for either a licence, environmental authorisation, environmental authority or development approval in order to undertake the waste disposal or recycling activity: EP Act (SA), s. 36 and Schedule 1; EMPC Act (Tas), s. 25 and Schedule 2; PEO Act (NSW), s. 48 and Schedule 1; EP Act (ACT), s. 42 and Schedules 1, 1A; and the WMPC Act (NT), s. s 30 and Schedule 2. Under the EP Act (Q), an environmentally relevant activity requires a development approval [EP Act (Q), Chapter 4, s. 73 and *Environmental Protection Regulation 1998 (Qld)*, Schedule 1].

All environmental harm-based jurisdictions have a comparable approach by linking the control of waste disposal and recycling to the statutory object of preventing environmental harm including adverse effects on human health: EP Act (SA), s. 10(b)(i); EP Act (Q), s. 4(5)(a)(iii) and *Environmental Protection (Waste Management) Regulation 2000 (Qld)*, s. 4; EMPC Act (Tas), s.109, Schedule 1, Part 2, para. 3(b); PEO Act (NSW), s. 3(d); EP Act (ACT), s. 2; and the WMPC Act (NT), s. 5.

Statutory considerations for licences and approvals

The relevant environmental protection statute sets out the matters (or criteria) that must be taken into account, as a legal obligation, in the administrative decision-making process undertaken by a State or Territory Government agency to determine whether to grant a licence, authority or developmental approval. These matters so become relevant considerations for the administrative decision-making process. A primary basis for a judicial review challenge for an administrative decision made in relation to the grant of a licence or authority is whether the actual matters or criteria prescribed in a specific statute for each jurisdiction have been taken into account in the administrative decision-making process. Some of the criteria or matters for the administrative decision-maker to consider in the harm-based jurisdictions include (i) the statutory objectives, (ii) ESD, (iii) environmental protection policies, and (iv) the potential of the activity to cause environmental harm including measures to protect against environmental harm, e.g. best practice environmental management: EP Act (SA), s. 47; EP Act (Q), s. 73A and Schedule 3, Standard Criteria; EMPC Act (Tas), s. 25; PEO Act (NSW), s. 45; EP Act (ACT), s. 61; and WMPC Act (NT), ss. 32, 33.

Conditions for approval

In addition, the Government agency, as part of the administrative decision it makes, has general powers to impose conditions on the licence, authorisation or approval: EP Act (SA), s. 45; EP Act (Q), s. 73B; EMPC Act (Tas), s. 25(6); PEO Act (NSW), s. 63; EP Act (ACT), s. 51; WMPC Act (NT), ss. 35, 36; EP Act (Vic), s. 21 and EP Act (WA), s. 62. There is a legal obligation for conditions to be reasonable and relevant to the development or use for which the licence is granted. Conditions must not be unclear in their application (Bates, 2006).

Statutory duties — public duties

Environmental protection legislation in some States and the Territories prescribes a number of statutory duties. An action to enforce the performance of a statutory duty by way of a judicial review requires the duty to be sufficiently specific, rather than being too vague, or of a general or undefined character (Wade and Forsyth, 2004). The most rigid form of a statutory duty is prescribed in Queensland [EP Act (Q), s. 5] which makes it mandatory for any person exercising a function or power under the Act, e.g. an administrative decision-maker granting an environmental authority, to exercise that power in the way that best achieves the object of the EP Act — environmental protection and ESD. Reasonably similar, but not quite as prescriptive, provisions are contained in the environmental protection legislation for South Australia [EP Act (SA), s. 10(2)]; Tasmania [EMPC Act (Tas), s. 8]; New South Wales [PEO Act (NSW), s.294A]; and the Australian Capital Territory [EP Act (ACT), s. 12(2)].

Additionally, a "*general environmental duty*" is prescribed in some States and the Territories which requires that an activity cannot be undertaken which harms, or has the potential to harm, the environment, unless all reasonable and practicable steps are taken to prevent or minimise environmental harm. Matters that must be taken into account, as a legal obligation for complying with the duty, include: the sensitivity of the receiving environment; current status of technical knowledge; financial implications; and likelihood of success of the proposed measures: EP Act (SA), s. 25; EP Act (Q), s. 319; EMPC Act (Tas), s. 23A; EP Act (ACT), s. 22 and WMPC Act (NT), s. 12. Whilst failure to comply with the general environmental duty, in itself, does not result in civil liability, compliance may be enforced by a civil remedy to restrain a breach of the legislation (Bates, 2006).

Statutory injunctions

Apart from the common law principles for a judicial review, jurisdictions having a harm-based approach to environmental regulatory control in Australia also provide a statutory basis for enforcement through civil proceedings: EP Act (SA), s. 104; EP Act (Q), s. 505; EMPC Act (Tas), s. 48; PEO Act (NSW), s. 252 and EP Act (ACT), s. 128. The statutory approach of the WMPC Act (NT) is for civil rights and remedies for civil enforcement proceedings for environmental harm to be preserved (s. 7).

Furthermore, in some harm-based jurisdictions, where a contravention may cause environmental harm, the use of civil enforcement proceedings extends so as to apply to any Act — other than only for the actual harm-based statute: EP Act (SA), s. 104; EMPC Act (Tas), s. 48 and PEO Act (NSW), s. 253.

(b) The United Kingdom

Waste management is based on EU Directives that have been implemented into domestic law.

Sources of law

The Council Directive 91/689/EEC *on hazardous waste* (the "Hazardous Waste Directive") represented a significant direction for regulatory control of hazardous waste. As well as setting out broad principles for effectively managing and monitoring hazardous waste, it provided an objective basis for the listing of wastes. Three annexes to the Directive have defined the characteristics of the waste that must be satisfied for the waste to be listed as a hazardous waste — based on the nature of the activity, its hazardous constituents and its hazardous properties. Provisions in the EU Directive for hazardous waste for the listing of hazardous wastes and for matters such as licensing were implemented in the *Hazardous Waste (England and Wales) Regulations 2005* (the "HW Regulations"). For the purpose of the HW Regulations, reference is made (at Reg. 2) to the conditions in Article 4 of the earlier Council Directive (Directive 75/442/EEC *on waste*). Specifically, conditions "*to ensure that waste is recovered or disposed of without endangering human health*" and without the use of processes or methods which harm the environment — including, for example, "*without risk to water, air, soil, plants and animals*."

Further EU Directives are more specific in that their focus is on methods of waste treatment. For example: Council Directive 96/61 *concerning integrated pollution prevention and control* (the "IPPC Directive"), provided for regulatory control of waste treatment at installations such as landfill sites and incinerators. The *Pollution Prevention and Control Act 1999* (the "PPC Act") and the *Pollution Prevention and Control (England and Wales) Regulations 2000* (the "PPC Regulations") implemented the IPPC Directive.

Characterisation of hazardous waste

The approach adopted in the EU/UK is reasonably analogous to the model applied in the *Resource Conservation and Recovery Act* in the United States. Under the HW Regulations, a hazardous waste is defined as either:

i. A listed waste (Reg. 6, HW Regulations);
ii. Waste that exhibits one or more hazardous characteristics specified in Schedule 3 of the HW Regulations (Reg. 8, HW Regulations); or
iii. Is a mixture of the two.

In addition, the term "hazardous waste", as defined in the EU Hazardous Waste Directive, now replaces what were termed "special wastes" under the *Special Waste Regulations 1996* (Reg. 74, HW Regulations). The following Schedules to the HW Regulations, linked to Annexes to the Hazardous Waste Directive, are relevant in characterising a hazardous waste:

- Schedule 1 HW Regulations (Annex IA, 1B to the Hazardous Waste Directive): The Schedule lists "categories of hazardous waste" according to the nature of the activity which generated them. The waste may be in the form of a solid liquid or sludge.
- Schedule 2 HW Regulations (Annex II to the Hazardous Waste Directive): The Schedule identifies the "constituents of the waste" (in Annex IB) which render them hazardous when they have the "properties" described in Annex III of the Hazardous Waste Directive. Schedule 3 (Annex III to the Hazardous Waste Directive) describes the properties of wastes which render them as hazardous, e.g. corrosive, flammable, toxic, carcinogenic, teratogenic, infectious or ecotoxic. Ecotoxic refers to "*substances and preparations that present or may present immediate or delayed risks for one or more sections of the environment*" (Annex III).

Where a safe limit (or threshold) value of concentration applies to a hazardous waste contained in the List of Wastes, it will only be considered as a hazardous waste where the relevant limit value of concentration is exceeded. Consequently, the potential for such a waste to endanger human health could be established with greater certainty relative to a waste where no limit value exists.

Waste management licensing

Waste management licensing of installations for landfill sites and incinerators depends on whether licensing comes under domestic law or an EU Directive. Whether licensing of a waste installation comes within the *Environmental Protection Act 1990* or the IPPC Directive/PPC Regulations can be determined by reference to the list of controlled installations and activities as prescribed in Schedule 1 of the PPC Regulations. For example, Schedule 1 of the PPC Regulations (Chapter 5, "Waste Management") describes *Disposal of Waste by Incineration* (Part 5.1) and *Disposal of Waste by Landfill* (Part 5.2), and sets out the matters that are to be taken into account, as a legal obligation, in the administrative decision-making process.

Licensing under the Environmental Protection Act

Waste installations that do not come within the description of activities under Schedule 1 of the PPC Regulations fall within the waste management licensing system of the *Environmental Protection Act*. Applications for a waste management licence are made to the Environment Agency under s. 36 of the *Environmental Protection Act*. The matters that are to be taken into account by the administrative decision-maker as a legal obligation in determining the application are prescribed in s. 36(3). These matters provide the administrative decision-maker with discretion in determining whether to grant permission to the application, with or without conditions. In considering an application the waste regulation authority must be satisfied that the applicant is a fit and proper person. In such a case, an application will not be rejected unless the authority is satisfied that its rejection is necessary for the purpose of preventing:

- Pollution of the environment [s. 36(3)(a)];
- Harm to human health [s. 36(3)(b)]; or
- Serious detriment to the amenities of the locality [s. 36(3)(c)]. However, the amenity provision is inapplicable where there is planning permission in force in relation to the use to which the land will be put under the licence.

Permits under the IPPC Directive

Permitting waste installations that come within the activities prescribed in Schedule 1 of the PPC Regulations is undertaken through the IPPC Directive. Under the IPPC Directive, control only applies to activities carried out for the most polluting industrial installations; many small and medium sized enterprises are excluded (Bell and McGillivray, 2006).

A permit under the IPPC Directive aims to either prevent or to reduce emissions through technical solutions. Control under the permit is based on the "best available technique" — by reducing emissions and the impact on the environment using techniques which allow implementation in the relevant industrial sector under economically and technically viable conditions: Article 11, IPPC Directive.

Statutory Duty and the PPC Regulations

Regulation 8 of the PPC Regulations prescribes the discharge and scope of functions exercisable by the Environment Agency. Specifically, "*for the purpose of achieving a high level of protection of the environment taken as a whole by, in particular, preventing or, where that is not practicable, reducing emissions into the air, water and land*": Reg. 8(2).

(c) The United States

Regulatory control of hazardous waste by the Federal Government arises under the *Resource Conservation and Recovery Act of 1980* (the "RCRA"), which amended the originally enacted *Solid Waste Disposal Act of 1976*: 42 U.S.C. Sections 6901 *et seq.* The RCRA is designed to protect human health and the environment from risks associated with hazardous wastes and the handling, storage, transportation and disposal of hazardous waste materials.

The meaning of hazardous waste

Section 1004 of the RCRA provides the following definition for hazardous waste:

> "(5) The term 'hazardous waste' means a solid waste, or combination of solid wastes, which because of its quantity, concentration, or physical, chemical, or infectious characteristics may —
>
> (A) cause, or significantly contribute to an increase in mortality or an increase in serious irreversible, or incapacitating reversible, illness; or
>
> (B) pose a substantial present or potential hazard to human health or the environment when improperly treated, stored, transported, or disposed of, or otherwise managed."

Characterisation of hazardous waste

Where a solid waste does not come within the description of an exclusion under 40 C.F.R., Section 261.4(b), there are various categories that the EPA may consider for a solid waste to be hazardous. These categories have been summarised by Plater *et al.* (2004) and Craig (2005):

- *Listed hazardous waste*: Wastes that satisfy the RCRA definition of a hazardous waste can be regulated if they have been determined by the EPA as hazardous and so become specifically listed as a hazardous waste in the EPA's regulations: 40 C.F.R., Section 261.3(a). However, not all hazardous wastes are included on the EPA list.
- *Characteristic hazardous waste*: Where there is evidence to show that a waste possesses one or more of the hazardous characteristics specified in 40 C.F.R., Section 261.20–24 (Subpart C) — specifically, the characteristics of ignitability, corrosivity, chemical reactivity or toxicity, such a waste will come under the RCRA as a "*characteristic waste*" and so be distinguished from a listed waste.
- In addition, where a listed waste is mixed with a non-hazardous waste or material the mixture must be managed as a hazardous waste. There are two exceptions to this: where hazardous debris satisfied Land Disposal Restriction standards and for residues from the processing of certain wastes that use high temperature metals recovery processing (US Department of Energy, 2007).
- A final category relates to a waste that is a result of the storage, disposal or treatment of hazardous waste (Plater *et al.*, 2004).

Hazardous waste disposal permits

The RCRA requires any facility that is treating, storing or disposing of hazardous wastes to obtain a permit: Section 3005 RCRA. Section 3005(c)(3) requires the administrative decision-maker to impose such terms and conditions on the permit as determined "*necessary to protect human health and the environment*". Whilst the RCRA (at Part III) provides for the EPA to be the permitting authority for toxic storage or disposal facilities, Part IV of the RCRA provides for the EPA's power to regulate toxic storage and disposal facilities to be delegated to the States. The RCRA provides the EPA with discretion to determine whether a State programme is acceptable. However, EPA Regulations provide the EPA with the right to set aside and to replace State permit conditions with Federal conditions in the State plan (Ferrey, 2004).

Exposure information and health assessments

From a community perspective, the following legal obligations imposed by the RCRA (at Section 3019) are particularly significant in relation to the object of the RCRA to protect human health and the granting of a permit for a hazardous waste facility (Environmental Law Institute, 2001, 2002):

- A requirement for each application for a permit to include information about "*reasonably foreseeable releases*" from normal operations, potential pathways and the magnitude of human exposure to releases [paragraph (a)];
- A health assessment may be requested where, in the judgement of the EPA (or a State where appropriate), a land disposal facility "*poses a substantial potential risk to human health*" [paragraph (b)]; and
- Priority for a health assessment is given to sites where "*potential risk to human health appears to be highest*" and where the existing health assessment data are inadequate to assess potential risk [paragraph (d)]. Health assessment is defined [at paragraph (f)] to incorporate matters such as "*potential for pathways of human exposure*" *and* "*potential susceptibility of the community within the likely pathways of exposure*".

Chemicals and the environment: other federal legislation

The Toxic Substances Control Act [15 U.S.C. Sections 2601 *et seq.* — the "ToSCA"] focuses on risks to human health and the environment from new and existing chemical substances. Regulatory control is achieved through approaches that screen new chemical substances, a permitting and approval process and by placing restrictions on substances that represent "unreasonable risks" for health and the environment. Under Section 5(b)(4)(A) of ToSCA, the EPA may compile and keep a list of chemical substances, the disposal of which may present an unreasonable risk of injury to health or the environment. In making such a finding, relevant factors to consider include the effects of the chemical substance on health or the environment — and the magnitude of human or environmental exposure to the chemical substance. Given that this provision is reasonably similar to some of the disposal permit requirements under the RCRA, it would be a prudent step to review information available under the ToSCA for a chemical substance where the chemical may be the substance in issue in a hazardous chemical–public health conflict.

INTERNATIONAL ENVIRONMENTAL CONVENTIONS AND GLOBAL CHEMICAL USE

Global trade in hazardous waste is a multi-billion dollar industry. Disposal of hazardous waste has become a profitable business for developing countries. However, because of the possibility of relaxed disposal regulations in some developing countries, concern has been expressed at the potential for increasing shipments of hazardous wastes for disposal in these countries (Vilcheck, 1991).

A key international agreement dealing with the global movement and disposal of hazardous wastes is the Basel Convention *for the Control of Transboundary Movements of Hazardous Wastes and their Disposal* (1989). By July 2004, the Convention had been ratified by 162 countries (Brunnée, 2004). The primary obligation under the Convention is for a Member State to ensure that any hazardous wastes which that State exports are not exported without the previous consent of the receiving countries. In addition, the safe transfer and disposal of waste must be ensured. Any imported or exported hazardous wastes must be managed in an "*environmentally sound manner*". The meaning of the concept "environmentally sound manner" is described in the Basel Convention Technical Guidelines (UNEP, 2005). Buccini (2004) considers that whilst the positive feature of the Basel Convention is that it represents risk management commitments, it is limited by its omission to incorporate other essential steps for assessing issues related to hazardous chemicals in the environment. Specifically, problem identification and priority-setting, risk assessment, and monitoring and evaluation.

The Basel Convention was ratified by Australia on 5 February 1992 and implemented into domestic law in the *Hazardous Waste (Regulation of Exports and Imports) Act 1996 (Cth)*. The United Kingdom ratified the Convention on 7 February 1994. By Council Regulation 259/93 (*on the supervision and control of shipments of waste within, into and out of the European Community*), the Basel Convention was implemented within the European Community. However, the position in the United States may be contrasted. The United States signed the Convention in 1989 and, notwithstanding that the Senate gave its advice and consent to ratification in 1992, the United States remains the only OECD country not to have ratified it (Brunnée, 2004). Brunnée notes that around 1 per cent of the hazardous waste produced in the United States each year is exported to Canada and Mexico, facilitated through the "North American Free Trade Agreement (1992)". This Agreement eliminated barriers to waste trade between the United States, Mexico and Canada (Rogus, 1996).

The Basel Convention and Customary International Law

Rogus (1996) contends that under customary international law, the liability a State may incur for acts of its private individuals or entities is directly related to that State's control over those acts, whether or not they occur in its territory. A consequence of ratifying the Basel Convention would be for Member States to exercise direct control over exports of hazardous waste. Rogus further contends that ratification of the Convention may, in all likelihood, increase a Member State's environmental regulatory control over the acts of its private individuals or entities and so it may be subject to liability under the general principles of international law — including the possibility of becoming a target for a law suit based on its regulatory control over its citizens who export hazardous waste. However, one qualification raised by Rogus is whether customary international law would have sufficient strength to prevent environmental damage and whether it provides sufficient sanctions directed against the perpetrators of the damage.

LEGAL RIGHTS AS A SOURCE OF KNOWLEDGE POWER FOR NEGOTIATING SOLUTIONS

Applications for a permit or a licence in the Australian States and Territories do not provide for effective public participation in the application process as any community consultation, in general, will be limited. A similar situation exists in the UK under the *Environmental Protection Act 1990*. In these circumstances, only the administrative decision to grant a permit or licence or authorise the use of the site may be the subject of a judicial review challenge. A potential judicial challenge would primarily be based on general grounds such as whether there had been an error of law in the administrative decision-making process; or whether there was no evidence to support the findings of fact made by the administrative decision-maker; or a decision that was not authorised by the statute under which the licence or permit was granted because of a failure to perform a public duty prescribed by the statute.

In the UK where the IPPC Directive applies for applications for new permits for new installations, or for substantial changes, the permitting procedure does require public comment before the administrative decision is made: Article 15, IPPC Directive. In the USA, the RCRA, at Section 7004(b)(1)(2), requires the agency to publicise in the media its intention to issue a permit for any facility to treat, store or dispose of hazardous waste before the administrative decision is made. Public hearings (including an opportunity for presentation of written and oral views) are held where written notice opposing the agency's intention to issue a permit is received and a hearing is requested — or where the agency, on its own initiative, determines to hold a public hearing. For the UK at least, under the IPPC Directive, the public comment procedures prescribed by legislation provide a further possible ground for judicial review in addition to the general grounds of review of a decision to grant a permit or licence. Procedural fairness may arise in the context of the element of legitimate expectation and full and fair disclosure of information during the public comment stage. The availability of such an option under the RCRA in the United States would depend on the law relating to procedural due process to have developed equivalent principles for legitimate expectation and full and fair information disclosure, equivalent to UK public law.

In the United States, grounds for a judicial review challenge for the administrative decision made under the permitting process for the RCRA may include a consideration of whether an error of law had been made in making the decision and whether the administrative decision could be substantiated in relation to the "arbitrary, capricious or abuse of discretion" test as well as the requirements for substantial evidence and the "hard look" doctrine.

HAZARDOUS CHEMICAL AND PUBLIC HEALTH CONFLICTS: A SCIENTIFIC PERSPECTIVE

Understanding the distinctive features of hazardous chemical exposure and public health problems represents the first step for providing a foundation of knowledge power to participate effectively in conflict resolution. Some of the distinctive features (Christie, 1992a,b) include:

i. In contrast to a whiplash injury arising from an automobile collision where there is a well-defined, direct link between the injury and the event, hazardous chemical exposure results in a genetic or biochemical disturbance, e.g. a cancer, where the link between exposure and causation of a disease is not always direct (Gold, 1986);
ii. The disease arising from hazardous chemical exposure is not always immediate. Whilst exposure to the chemical may increase the risk of contracting a disease, exposure does not produce an immediately observable response. A long lag time (or "*latency period*") between exposure to the chemical and the emergence of symptoms often exists and may take up to 20 years, or more (Wagner, 1986);
iii. Exposure to hazardous chemicals may involve acute exposure, i.e. exposure to a high chemical concentration arising from a single

catastrophic event such as the nuclear accident in 1986 at the Chernobyl Nuclear Power Plant in Ukraine of the then USSR. Alternatively, there may be chronic exposure, i.e. exposure to low chemical concentrations, intermittently, over long periods of time;

iv. Detecting and establishing adverse effects of chronic exposure to chemicals are difficult as the onset of the disease and observable symptoms will be gradual. For chronic toxicity, the end-point requires an estimate of the integrated lifetime exposure. In contrast an effect may be observed from a single instance following an acute exposure (Severn, 1987);

v. Where there is a long latency period before observable symptoms emerge, there may be a degree of uncertainty in recalling the characteristics of exposure to the hazardous chemical, throughout the duration of exposure. There are only a small number of examples where the link between exposure to a specific chemical and causation of a disease is quite clear;

vi. These examples involve very rare tumours noted on disparate populations linked only by exposure to a common chemical (Brennan and Carter, 1985). They have been referred to as "*signature diseases*", the disease leaving the signature of the substance (Farber, 1987). Notable examples include hepatic angiosarcoma in PVC workers, clear cell adenocarcinoma of the vagina and cervix in women exposed to DES (diethylstilboestrol) in utero and mesothelioma in asbestos workers; and

vii. In contrast to signature diseases are the more general health complaints related to chemical exposure. These claims may have difficulty meeting the legal and scientific requirements for establishing connection between exposure and the health problem. For example, claims that exposure to a chemical has caused, or may cause, health problems such as suppression of the immune system or complaints typically having non-distinctive or non-specific symptoms, such as headaches, skin irritation, allergy and breathing difficulties (Christie, 2005).

HAZARDOUS CHEMICAL AND PUBLIC HEALTH LITIGATION: A LEGAL PERSPECTIVE

Understanding the distinctive features of a hazardous chemical–public health problem is central to considering pathways for resolving conflict and for undertaking a BATNA analysis — including whether there is an adequate basis of power to effectively participate in negotiating agreement.

The following analysis of past hazardous chemical–public health litigation compares one successful example (asbestos litigation) with an unsuccessful case ["Agent Orange" (Dioxin) litigation] when both the asbestos conflict was considered by courts in the United States (Ginsburg, 1986; Schuck, 1987) and the dioxin conflict was the subject of an inquiry in Australia ("*Royal Commission on the Use and Effects of Chemical Agents on Australian Personnel in Vietnam*", 1985). The analysis identifies the key elements that needed to be established by the scientific and medical evidence. All of the following key elements were satisfied by factual evidence in the asbestos litigation:

i. The long latency period for asbestos exposure and mesothelioma had run its course by the time the case had reached trial so that the toxic effects of the chemical had been fully investigated and revealed;
ii. The disease occurring after the latency period was specific for exposure to asbestos. The two most common asbestos-related diseases, asbestosis and mesothelioma, were relatively exposure- and disease-specific. Both diseases were rarely, if ever, seen in persons not exposed to asbestos;
iii. The levels of asbestos exposure were relatively high and sustained. It could be established what levels of exposure had occurred, as well as whether a person had actually been exposed to asbestos; and
iv. The symptoms for asbestosis and mesothelioma were distinctive and, ultimately, easily observed.

In contrast, at the time of the "Agent Orange (Dioxin)" Royal Commission in Australia, (i) the latency period for dioxin exposure and disease specificity had not run its course, (ii) the available epidemiological studies were characterised by high statistical variability, (iii) some of the diseases complained were considered to have other possible sources of causation and not dioxin-specific and (iv) there were concerns as to the weight that could be attached to the exposure data.

(a) A Judicial Analysis of Objective Scientific Evidentiary Needs

Muller v Queensland Electricity Commission [1999] QSC 123 was an Australian common law negligence litigation case, decided by the Supreme Court of Queensland, involving isocyanate exposure. In his reasons for decision, Shepherdson J endorsed a number of published medico-legal research articles on hazardous chemical litigation in making the following observations on the factual evidence required to establish whether exposure to a hazardous chemical may have caused the medical condition in issue:

i. The nature and duration of exposure to a particular chemical must be characterised — a problem that may be exacerbated, not only where there is a long latency period, but also where there is a possibility that exposure to other chemicals may also have occurred over time;

ii. Scientific research published in the literature undertaken independently of the parties to the litigation can be applied in order to establish whether a particular chemical can cause the medical condition in issue;
iii. The particular chemical to which a party may have been exposed, or was exposed, must be the same as that in the published research study. Also, the potential or actual exposure history that is being claimed in relation to dose (or concentration) and duration, must approximate that of the subjects in any published research study relied upon; and
iv. Evidence relied upon must distinguish between a disease allegedly caused by exposure to the specific chemical and similar diseases that may be induced by the general environment. Also, the disease should be specific for exposure to the particular chemical that is the subject of the claim.

(b) Changing Scientific Knowledge and Causation in Hazardous Chemical–Public Health Conflicts

The relevant scientific and medical knowledge for resolving a chemical exposure–public health conflict, in common law negligence litigation, is the existing knowledge at the time of potential or actual exposure to the chemical — and not at the time of disease onset. There must be an accepted body of knowledge from which it could be established that it would be reasonably foreseeable that exposure to the hazardous chemical would result in a specific public health problem. This is a very significant factor to recognise in any BATNA analysis, particularly where there is a long latency period between exposure to the chemical and the emergence of symptoms.

In the landmark asbestos case successfully litigated in the United States, *Borel v Fibreboard Paper Products Corporation* 493 F.2d 1076 (1973), the United States Court of Appeal, Fifth Circuit, reviewed the published scientific literature over the working life of Clarence Borel with the Fibreboard Paper Products Corporation, from 1936 (the year he commenced work in industrial installation) to 1969 (the year he was first diagnosed with asbestosis). The court concluded that the danger from inhaling asbestos dust by insulation workers was foreseeable during his working life. However, the first warnings were not given until 1964–66, by which time it was too late for him.

PROOF OF CAUSATION IN HAZARDOUS CHEMICAL–PUBLIC HEALTH CONFLICTS: MEDICO-LEGAL PRINCIPLES

The central issue for proof of causation in a hazardous chemical–public health conflict is the need to establish a connection between the chemical, exposure and disease. This requires both proof of legal causation and proof of scientific causation to be established.

(a) Proof of Legal Causation

Proof of legal causation refers to factual proof that the conduct of another party, through an act or omission (i.e. a failure to act), may have resulted in, or have the potential for, the escape or release of a chemical into the environment, leading to an actual or potential public health problem. The effectiveness of an environmental management plan or the monitoring programme at the site, as well as the extent of adherence to the principles for due diligence, are all matters that come within the scope of proof of legal causation. In the context of hazardous chemicals–public health conflicts, common law principles, arising from the common law, complement statutory rights contained in environmental protection legislation. Together, they represent a significant source of knowledge power as they are the foundation for effectively participating in negotiating agreements.

(b) Proof of Scientific/Medical Causation

Proof of scientific/medical causation refers to the scientific and medical evidence that establishes whether exposure to the specific chemical has caused, or may have the potential to cause, a public health problem. Scientific data and information need to address the following matters:

i. Scientific/medical research and publication into the association between exposure to the chemical and the disease incidence;
ii. The characteristics of exposure to the chemical; and
iii. Characterisation of the risk for contracting the disease following exposure to the chemical.

SCIENTIFIC CAUSATION: RESEARCH INVESTIGATIONS

Questions of scientific causation in hazardous chemical–public health conflicts may not be able to be evaluated directly for a number of reasons, the most important being:

i. That the persons exposed (or potentially exposed) to the chemical may never have been the subject of a published experimental study that investigated the effects of the specific chemical on their health; and
ii. That the conflict has arisen where there may be a potential situation for an escape of a hazardous chemical into the environment that may adversely affect public health — but the event has not occurred.

The approach used by science and law to assess facts and questions of causation, indirectly, is by the method of inference. An inference requires there to be objective facts, from which to infer the other facts to establish scientific causation. Where there are no such positive, proven facts from which an inference can be made, the method of inference fails and what is left is mere speculation or conjecture: *Caswell v Powell Duffryn Associated Collieries* [1939] 3 All ER 722.

In order to establish proof of scientific causation, indirectly by the method of inference, reliance must be placed on published research involving both laboratory animal studies and epidemiological studies with human populations — notwithstanding that the community who may be potentially exposed to the chemical may not be the subject of the published epidemiological studies. The published studies in the scientific literature relied upon must represent an accepted body of scientific knowledge. To achieve this goal, the scientific information needs to address the following matters:

- Whether the specific chemical has been shown to be clinically hazardous to test animals under strictly controlled laboratory conditions; and
- Whether epidemiological studies show a statistically significant correlation (or "association") between exposure to the chemical and changes in the incidence of disease in a human population.

However, science and law both recognise that where the relationship between exposure to a chemical and incidence of disease has been found to be statistically significant, a mere correlation, by itself, does not establish cause and effect. Therefore, the epidemiological data must be evaluated for biological plausibility. One approach to demonstrate biological plausibility, accepted by epidemiologists, is through the use of the "Henle–Koch–Evans Postulates" (Evans, 1976). Epidemiology complemented by animal studies, together with an objective evaluation of findings from these studies in terms of biological plausibility, makes it possible to strengthen an inference that (i) a chemical shown to be clinically hazardous to test animals under strictly controlled laboratory conditions and (ii) shown, epidemiologically, to be correlated with changes in the disease incidence in human populations, may have caused, or may have the potential to cause, a public health problem.

(a) Animal Studies: Extrapolation from Animals to Humans

Assessing the potential risks of chemicals to humans, by extrapolating experimental findings on laboratory animals to possible health effects on humans, is a step associated with some debate. Because of major biological differences between species in terms of physiology, metabolic differences and anatomy, routes of administration and exposure regimes (e.g. the dose rates used), it is not surprising that scientific debate exists as to the validity of extrapolation (Schwartzbauer and Shindell, 1988).

Whilst there is no perfect surrogate for humans for evaluating toxicities of chemicals, and given the inherent limitations of laboratory animal studies, it is reasoned that they remain the best model for evaluating carcinogenic potential and other toxicities of chemicals (Office of Technology Assessment Task Force, 1988). Although laboratory animal studies may not be able to predict absolute risk for humans, their use may indicate which chemicals might be of greater concern to human health than others (Rall, 1988). Toxicologists would argue that, notwithstanding their limitations, it is prudent to accept animal studies as a means for predicting effects of chemicals on humans and in determining when risk quantifications should proceed (Brennan and Carter, 1985).

(b) Epidemiological Studies: Extrapolation to Individuals

Epidemiology is an observational science that involves the study of the pattern of disease occurrence in human populations, not individuals, and identifies the factors that influence these patterns. In contrast to animal studies, epidemiology represents the only means of directly evaluating the potential impacts on human health from chemicals. The primary objective of epidemiological research is to determine the incidence of disease and to calculate the risk to others of developing the same disease (Ginzburg, 1986). Epidemiological studies apply statistical techniques and probability theory in order to identify and to establish the cause of known human disease (Black and Lilienfield, 1984).

The object of statistical analysis in epidemiology is to determine whether there is a statistically significant difference in the incidence of disease between people who have been exposed to a hazardous chemical and people who have not been exposed to it. Statistical significance means that the differences in the incidence of disease observed between the two groups does not occur by chance, but rather through exposure to a particular hazardous chemical. The final step in an epidemiological study is quite independent of statistics. It requires a biologically plausible inference to be drawn between the relationship of a specific hazardous chemical to the disease "aetiology", i.e. the specific agent causing the disease (Black and Lilienfield, 1984). There are two specific features of epidemiological studies that must be recognised:

- Although epidemiological studies can predict what will happen to an entire population following exposure to a specific hazardous chemical, an epidemiological study is unable to predict which individuals will develop disease following exposure to that chemical; and
- There are far more sources of experimental variability in epidemiological studies, compared with animal studies. The degree of control adhered to in animal studies to minimise experimental variability cannot be achieved in epidemiological studies, e.g. uniformity in the type and duration of exposure of the human population — including control over the dose rate of the chemical each person receives. Variability in the experimental data acquired in the study will be reflected in the confidence interval around the estimates of risk and so affect the level of significance that can be attached to the experimental findings. The extent of variability is crucial for determining the weight to attach to the results of an epidemiological study.

Types of epidemiological study

There are a number of approaches, but epidemiological output is dominated by two types of study: the *case control (or retrospective) study* and the *cohort (or prospective) study* (Paddle, 1990). Both the case control and the cohort study involve the use of a "control" group — a group of people who have not been exposed to the hazardous chemical. For both studies, in order to minimise experimental variability, both groups need to be identical in all respects, as far as possible — apart from their exposure history. The purpose of each study is to determine whether any difference in disease incidence can be attributed to exposure history.

Case control studies A case control study is a retrospective study in which the disease has already occurred. The technique involves the selection of a group of diseased cases, finding a comparable group of people who have not been exposed to the hazardous chemical, and then investigating the exposures within each group.

Cohort studies A cohort study is a prospective study. It involves finding a group of people that have been exposed to a chemical, and a comparable group of people that have not been exposed to the hazardous chemical, and to then investigate the occurrence of disease, over time, within each group.

Case control v cohort studies The case control study is the most frequently used epidemiological study and is the approach most likely to be offered in evidence in hazardous chemical litigation. A primary area of scientific concern of case control studies relates to their reliability because of the increased possibility of both subject and observer bias. Subject bias could arise if an epidemiological study is undertaken as the basis for litigation or negotiation, as the exposed group could well provide biased answers which would favourably influence the outcome of litigation or negotiation. Because of the latency issue for disease onset, bias may also arise where the exposure history of individuals in the exposed group is based on independent recall or memory (Andrues, 1989).

Some features in the cohort study may also limit its wider use. For example, it is not always possible to identify populations that have been exposed or not exposed to the hazardous chemical. Also, the long latency period for disease onset frequently results in the epidemiologist being unable to follow both the exposed and non-exposed group for the time required (Black and Lilienfield, 1984).

SCIENTIFIC CAUSATION: EXPOSURE ASSESSMENT DETAILS

The question of information on exposure assessment is equally as important as research studies for resolving a chemical exposure–public health conflict. An effective exposure assessment requires the chemical hazard to be characterised in terms of the following objective data: (based on Severn, 1987; Bell, 1988; Andrues, 1989; Buccini, 2004; Tynan, 2005):

i. A history of the site including what hazardous chemicals would be present on site, e.g. a listing of chemical names, chemical structures and properties. Any records of background levels of the chemicals at the site should be obtained. Freedom of Information legislation may apply if measurements have been made by a Government agency;
ii. Identification of the potential sources for the release of the chemical which might result in the chemical ultimately coming into contact with the public. In assessing the sources of risk for the potential release of a chemical, the widest range of activities should be considered, including:

- Handling and transportation;
- Accidents relating to transport and manufacturing;
- The use of the chemical; and

- Disposal of wastes from manufacturing processes and from the end-of-life stages of products;

iii. The behaviour of the chemical following its release needs to be characterised as the potential environmental fate and pathways following the release of a chemical are distinctive for each substance. For example: whether transport and distribution of the chemical occurs in different media — air, soil, sediment, water or biota; whether the chemical might be broadly distributed throughout all media or concentrated preferentially in only one medium; whether transport occurs over short or long distances as a result of natural processes in air and/or water; whether the chemical persists in the environment at the same concentration following release; whether it undergoes chemical transformations and degrades into other chemicals, or becomes diluted in concentration, as it is transported through the environment;

v. In the absence of any measured data obtained through experimentation or monitoring, potential changes in chemical concentration transported through the environment from the source or release site, reliance would have to be placed on expert opinion evidence;

vi. The characteristics of the population potentially exposed to the chemical should be identified. This should include information on their proximity to the source of the chemical as well as the possible duration and characteristics of exposure; and

vii. Details on the characteristics of the potential risk for exposure of the population to the chemical, including:

- Whether the potential risk of exposure from the source to the person was direct — or indirect, i.e. with the chemical moving through an intermediate pathway to the person such as in air, water or other media;
- Whether the potential risk of exposure would be acute or intermittent; and
- The manner of human exposure to the chemical, e.g. ingestion, inhalation or dermal.

SCIENTIFIC CAUSATION: CHARACTERISATION OF RISK

In the context of a hazardous chemical–public health conflict there are two different scientific research methods used to characterise risk of disease. Epidemiological analysis is expressed in terms of risk ratios in order to avoid any challenge that the findings are speculative and so have limited value for establishing causation (Hoffman, 1984). Health risk assessment methodology (as discussed in Chapter 7) is the alternative methodology.

(a) Epidemiological Studies and Risk Ratios

A number of statistical relationships for risk are used in epidemiological studies to interpret the magnitude of the association between exposure and disease incidence. Two such accepted measures, in common use, are the relative risk and the attributable risk.

In epidemiology, relative risk is used as a measure of the probability of developing a disease following exposure to a particular chemical. Relative risk is expressed in the form of a ratio: - "*The incidence rate of disease in the group exposed to the hazardous chemical: The incidence rate of disease in the control group not exposed to the hazardous chemical*". The relative risk will be 1 where the incidence rate for a disease is similar in both exposed and non-exposed groups. Where the relative risk is greater than 1, the incidence rate for a specific disease will be will be greater in the group exposed to the hazardous chemical compared with the non-exposed group.

The relative risk for the association between asbestos exposure and the incidence of mesothelioma was 50–80 (Black and Lilienfield, 1984). The relative risk for the incidence of lung cancer for smokers, relative to non-smokers, was 10 (Doll and Peto, 1981). However, because many diseases have multi-factorial agents that may cause the disease, where more than one factor may be the cause of disease, relative risk will rarely exceed 10. Whilst a statistically significant relationship is unlikely to be due to chance, the question of demonstrating a true hazard to human health, or whether chance is the explanation, may sometimes be best answered by a repetition of circumstances and observations (Hill, 1965). Therefore, consistency from study to study is a very important criterion for scientific acceptance of epidemiological findings because it would be improbable for a spurious association to emerge from several published studies reporting reasonably constant values for relative risk. This would be particularly the case if the epidemiological studies were undertaken by different researchers, on different human groups, and their experimental methodology was not in issue.

A limitation for the case control study is that the disease incidence rates amongst exposed and non-exposed groups cannot be directly determined. Consequently, relative risk cannot be directly estimated. However, a statistic, the *odds ratio* can be derived which approximates the relative risk in circumstances in which the disease incidence in the non-exposed group is low (Black and Lilienfield, 1984).

Attributable risk is also used as a statistical measure of association in complex situations when multiple factors are involved with the incidence of disease. In the context of multiple factors causing a public health problem, attributable risk could be estimated as a measure of the impact of a particular chemical exposure on the occurrence of a specific disease. It is defined as the proportion of a disease, in a group that is exposed to a particular hazardous chemical, which can be statistically attributable to that factor.

(b) Health Risk Assessment Methodology and Acceptability of Risk

In the context of a chemical exposure–public health conflict, the first three stages ("Hazard Identification", "Dose–Response Relationships" and "Exposure Assessment") are reasonably similar to the epidemiological approach. However, the final step, "Risk Characterisation", diverges. It relies on a mathematical model to predict the probability of the risk of a public health problem arising along the potential or actual exposure pathways of the hazardous chemical. The predicted probability of risk is then considered in terms of whether it represents an acceptable level of risk. Risk management measures are evaluated in order to determine whether the hazardous chemical risks can be managed to an acceptable level of risk in order to protect human health.

(c) Risk Characterization and Biological Plausibility: The Henle–Koch–Evans Postulates

The limitations of solely relying on a statistically significant relationship for establishing causation in chemical exposure–public health conflicts has long been recognised by science. Biological plausibility in the statistical relationship must also be established. Law also recognises plausibility to be a relevant consideration for satisfying the civil standard of proof. For example, in *Briginshaw v. Briginshaw* (1938) 60 CLR 336 at 361, a decision of the High Court of Australia, Dixon J (as he then was), commenting on evidentiary proof of causation based on the balance of probabilities, stated:

> " … when the law requires the proof of any fact, the tribunal must feel an actual persuasion of its occurrence or existence before it can be found. It cannot be found as a result of a mere mechanical comparison of probabilities independently of any belief in its reality."

A widely accepted approach used by epidemiologists to demonstrate biological plausibility is through the use of the Henle–Koch–Evans Postulates. These Postulates provide a means for determining whether the statistical relationships established in epidemiological studies — or the probability of risk predicted in a health risk assessment — make biological sense. Irrespective of the statistical significance of the epidemiological studies, where epidemiological or health risk assessment data fail to satisfy one of the Henle–Koch–Evans Postulates, the biological plausibility is weakened so that any conclusions as to causality are made difficult (Evans,

1976). The ten factors comprising the Henle–Koch–Evans Postulates include:

- "The prevalence rate of the disease should be significantly higher in exposed versus non-exposed populations;
- Exposure to the causal agent should be greater in diseased versus control populations where all other risk factors are held constant;
- Responses to exposure should follow a biologic gradient from mild to severe;
- Replication of the disease should occur more often in exposed versus non-exposed populations; and
- Elimination of the causative agent should decrease the incidence of the disease".

POWER AND TACTICS IN HAZARDOUS CHEMICAL–PUBLIC HEALTH CONFLICTS

Tactics used effectively are a major source of power in any negotiation. A tactic not identified from the outset and unable to be effectively countered during negotiations may result in a situation where there is an inadequate basis of power to effectively participate. In addition, the fairness in any negotiated agreement could be compromised. Some of the more obvious tactics that are likely to be used in a hazardous chemical–public health conflict include:

- That it would be unlikely for a public health problem to arise based on the concentration of the chemical and/or the potential or actual exposure characteristics for the site;
- That the particular chemical could not cause a public health problem. Another explanation for causation is provided, e.g. the disease is not exposure-specific for that chemical. Following on to this tactic, a "laundry list" of other possible causes may be advanced (Royal, 1987); and
- Toxic dumps are often characterised by the number of different chemicals deposited over time. Any of the chemicals may be the aetiological agent for a potential public health problem. Where a number of different chemicals are deposited at a site, they may intermingle over time and undergo chemical transformations so that the potential impacts on public health, at best, may be poorly understood (Andrues, 1989).

HAZARDOUS CHEMICAL SITING DECISION-MAKING: THE SCOPE FOR ADR AND INTEREST-BASED NEGOTIATION

Proposals for locating hazardous waste disposal sites or facilities for treating, using or recycling hazardous chemicals create awkward, difficult and controversial problems for administrative decision-making by Government agencies. These decisions are sustainable development issues as they all require a balancing of public health and environment, economic, and social gains and losses. Compensation has been recognised in the United States as being unsuccessful as a basis for siting a hazardous waste facility where the community's reluctance to the proposed site relates to adverse health impacts (Gerrard, 1994).

The traditional siting process used by Government agencies was first described by Dennis Dusick as the "*Decide–Announce–Defend Process*" (as cited by Susskind, 1985). The alternative approach is a cross-disciplinary one in which the aim of ADR and interest-based negotiation is to find a solution, by consensus, for those who have to live with the outcome of the siting decision.

(a) The Decide–Announce–Defend Process

The key steps and features of this process can be summarised as follows (Susskind, 1985; Gerrard, 1994):

- A Government agency seeks a mandate to investigate the need for a new or expanded hazardous treatment or disposal facility. A committee or a particular body is then assigned the task of proposing a solution, or initiating a process, for selecting an appropriate site. Whilst technical experts are engaged to assist the committee, public inputs reflecting their needs and concerns are minimal;
- Where technical siting criteria are used to select the objectively "best" site, the technological solution may well be counter-productive;
- A public educational campaign may be initiated with the purpose of justifying the choice of site;
- The public are provided with opportunities to participate in the siting process. A community reference or focus group may be appointed to advise the Government agency. Public hearings may be held. Public comment may be sought on consultancy reports dealing with siting issues;
- The power available to the community to participate in the siting process is generally only in the form of making recommendations to the agency.

Power does not extend to shared involvement and responsibility for decision-making in the siting process. The Government agency has the liberty to make or adopt whatever recommendations it considers appropriate; and

- In the worst case scenario, and after vigorously defending the logic of its actions, the decision by the Government agency may be challenged in a legal action. If the opponents are successful in blocking construction through litigation, the entire siting process begins again.

(b) An Alternative to the Decide–Announce–Defend Process: ADR and Interest-Based Negotiation

The effectiveness of negotiation for resolving hazardous waste siting conflicts is limited where there is an imbalance of power for community opponents (Gerrard, 1994). In these circumstances, Gerrard reasons that negotiation may do little to address the underlying causes of public opposition, such as threats to public health, as the community may be unable to extract the necessary concessions. However, a cross-disciplinary approach based on concepts from law, science and ADR may effectively offset the concerns raised by Gerrard for using negotiation to resolve a hazardous waste siting conflict, i.e. to provide the community with an adequate basis to share power with the Government agency and to effectively participate in finding solutions by mutual agreement. In these circumstances ADR and interest-based negotiation may be seen as an alternative approach to the Decide–Announce–Defend Process, particularly where community concern has become galvanised or ignited over potential public health problems from exposure to chemicals from the proposed site. The following concepts are relevant for using ADR and negotiation for finding solutions to resolve a hazardous chemical siting conflict (Susskind, 1985; Buccini 2004; Susskind and Hoben, 2004):

Problem identification and prioritisation
The first step would be to identify any connection between potential public health issues arising from exposure from the chemicals present at the proposed site. This step requires the review and analysis of scientific information from a wide range of sources — from information published in the scientific literature through to information held by Government that can be obtained under FOI legislation.

Characterisation of risk at the proposed site using a joint fact-finding approach
The risk associated with the potential hazardous chemical–public health problem needs to be characterised. Characterisation of risk requires an

assessment of risk in terms of a "factual" dimension. Objective science based on medical principles, e.g. the risk ratio from an epidemiological study or from scientific principles such as health risk assessment, is one aspect; a further consideration is a "socio-cultural" dimension of risk, i.e. values and public perceptions of risk held by the community (International Risk Governance Council, 2005). Where there is agreement that some form of risk characterisation study needs to be undertaken, a joint fact-finding approach should be adopted. Support for the findings of any such study commissioned will be facilitated if all sides could agree on the terms of reference for the study, the experts who would undertake the study and the scientific methodology for the study.

Risk reduction and management

Where the hazardous chemical risk at the proposed site can be shown to represent an unacceptable risk to public health, management of the risk must be considered. Risk management provides the basis for determining the appropriate measures required to control or manage the risk to acceptable levels. Given that the community has to live with the outcome of any siting decision, a process in which the responsibility for determining an acceptable level of risk is shared with the community authenticates community concerns, on what is ultimately a cardinal issue for resolving siting conflicts. Risk management requires a conflict resolution process that provides equal room to consider both the objective scientific evidence and any community value-based perceptions of risk (International Risk Governance Council, 2005). ADR and interest based-negotiation provides for such a dual approach.

Shared responsibility for monitoring and evaluation of the risk management measures

There is a paramount need for shared responsibility, using a joint fact-finding strategy in preparing a monitoring programme, an environmental management for the site or for undertaking a risk characterisation study. The procedures for monitoring hazardous chemical risk at the proposed site are crucial — not only for evaluating the effectiveness of the risk management measures to be implemented, but also for seeking concessions from those who have to live with the siting decision. Should monitoring indicate that risk management measures are inadequate or ineffective, new issues for public health may be identified which, in turn, must be resolved. As Susskind (1985) points out, "*no other strategy offers a more telling acknowledgement of the legitimacy of local concerns*"; those who have to live with the siting decision know they can trust the monitoring programme. Where a risk management measure is found to be inadequate and the community is empowered to have the problem remediated — even to halt operations at the site, without having to seek a court order, Susskind reasons that, in these circumstances, there would less likely be opposition to the siting decision in the first place.

Enforceability of obligations to mitigate adverse impacts
A number of approaches may be adopted for achieving this principle. The first is by setting a substantial bond. However, there may be a degree of scientific uncertainty in accurately predicting all unexpected adverse impacts on public health from the outset. Therefore, the use of the "flexible environment assurance bond" (Costanza and Perrings, 1990) may warrant consideration. Another means, is for all parties to agree to an independent scientific expert to periodically review environmental management of the hazardous facility in terms of adherence to agreed obligations arrived at by negotiation and for the findings to be published as a matter of public interest. Finally, a negotiated solution may require that the stages for the construction and operation of the site are phased, i.e. the "step-by-step" principle, so that the later stages only proceed if earlier obligations have been met.

FINDING SOLUTIONS FOR HAZARDOUS CHEMICAL–PUBLIC HEALTH CONFLICTS: LITIGATION V NEGOTIATION

Proposals for locating hazardous waste disposal sites or facilities are land use conflicts in sustainable development. They require a balancing of issues in public health and the environment, with economic and social gains and losses. For hazardous chemical–public health conflicts, this will also involve a consideration of risk in terms of social objectives such as social well-being and equity, economic objectives such as best practice technology and environmental objectives such as public health. Camacho (2005) considers that courts have limitations for determining the most appropriate outcome in a typical land use conflict where preferences have to be traded together and where there is a need to balance public policy concerns that affect the competing interests of multiple parties; these issues are central to resolving a hazardous waste siting conflict.

There are many significant advantages for using a collaborative approach to joint fact-finding and problem-solving, through ADR and interest-based negotiation for finding solutions for sustainability and siting conflicts. Solutions agreed to may be prospective, by anticipating future problems over hazardous chemicals. Also, solutions to address immediate concerns associated with hazardous chemicals and public health and the environment may be arrived at by mutual agreement. Moreover, ADR and interest-based negotiation provides the community with a process which is consistent with the "*obligation to endure*", as identified by Rachel Carson (1962). The community is being asked to assume the risks by the Government agency. However, ADR and interest-based negotiation provides the community with a role in the decision-making process with all information and facts in their possession. A collaborative approach to joint fact-finding and problem-solving is not part of the adjudicative or determinative processes conducted in the courts. Litigation is not a dispute resolution process as conflicts are not resolved (Street, 2005). Nor are the legitimate needs and concerns of those involved in a hazardous chemical siting conflict reconciled as the court does not resolve the underlying conflict. Consequently, these issues may re-emerge, in time, leading to further conflict.

Where ADR and interest-based negotiation is used as the process for resolving a siting conflict, those in the community who have to live with the outcome of the siting decision, scientific experts and the Government agency can be involved in deciding an acceptable level of risk based on the two accepted dimensions for the interpretation of risk: a "factual" dimension that consists of scientific outcomes which can be measured and a "socio-cultural" dimension that reflects how a particular risk is perceived by the community in terms of values and emotions (International Risk Governance Council, 2005). Finally, there is far more flexibility in negotiation, relative to litigation, to address changing scientific knowledge over time. In particular, the link between exposure to a specific chemical and a public health problem and its subsequent incorporation into a revised environmental management plan for the site.

9. Biodiversity and threatened species

"[After some 20 years of the operation of the Endangered Species Act of 1973 in the United States, with 382 recovery plans having been put in place for listed species] only five of the listed species have recovered ... some two hundred species will probably never recover ... the process of listing species has accelerated, but the other crucial elements in the program ... the designation of critical habitat and the development and implementation of a recovery plan, have not kept pace with listings."

Hardy (1992)

"[The low recovery rate for endangered species in the United States] can be largely attributed to the fact that ... the habitat and biological systems critical for their survival have become severely fragmented and destroyed by human intrusions."

Holst (1992)

"A new scientific truth does not triumph by convincing its opponents and making them see the light, but rather because the opponents eventually die, and a new generation grows up that is familiar with it." (trans. Frank Gaynor, 1950)

Max Planck 1858–1947

INTRODUCTION

Scientific concerns over the impacts of human use activities on the conservation of biodiversity have been addressed, over time, through the implementation of environmental legislation for the regulatory control of threatened species and biodiversity. A key statute for the environmental regulatory control of threatened species was the *Endangered Species Act of 1973*, introduced in the United States. Amongst its purposes was "*to provide the means whereby the ecosystems upon which endangered species and threatened species depend may be conserved*" [at Section 1531(b)]. This statute became a model for the introduction of endangered species legislation in other countries. In addition to domestic environmental legislation as the basis for the regulatory control of threatened species, multilateral international environmental conventions (or treaties) are also significant. The multilateral "*Convention on Biological Diversity*" (the "Biodiversity Convention"), signed at the United Nations Conference on Environment and Development at Rio de Janeiro in June 1992, represented a major direction taken by the global community towards the conservation of biodiversity. The Biodiversity Convention recognised the global dimension of the problem and the need for international cooperation to find solutions for the conservation of biodiversity.

Global recognition for the continuing need for the conservation of biodiversity is highlighted by the release, over time, of the World Conservation Union's (IUCN's) "*Red List of Threatened Species*". The IUCN Red List released in May 2006, the "2006 IUCN Red List", has identified a clear trend for an ongoing decline of the earth's biodiversity and provided the following summary of species threatened with extinction (IUCN, 2006):

- "16,119 species are threatened with extinction, but this may be a gross underestimate because fewer than 3% of the world's 1.9 million described species have been assessed by the Red List.
- Animals: 7,725
- In major species groups, the percentage of threatened species ranges between 12% and 52%
- Birds: 12% or 1 in 8
- Mammals: 23% or 1 in 4 are threatened
- Amphibians: 32% or 1 in 3
- Turtles and tortoises: approximately 42%.
- Conifers: 25% or 1 in 4
- Cycads (an ancient group of plants): 52%."

The 2006 IUCN Red List is particularly significant as it identified the following threats to biodiversity that must be addressed if effective solutions for resolving biodiversity conflicts are to be produced:

i. Humans are the main cause of extinction and the principal threat to a species at risk of extinction;
ii. The primary threats for the ongoing decline of biodiversity are habitat loss and degradation. These threats affect 86 per cent of all threatened birds, 86 per cent of the mammals assessed and 88 per cent of the threatened amphibians;
iii. Other threats contributing to biodiversity decline are the introduction of alien species, over-exploitation of natural resources, and pollution and disease; and
iv. An awareness of the significance of human-induced climate change as a serious threat to biodiversity now also needs to be recognised.

Moreover, he release of the IUCN Red List in September 2007 revealed that one in four mammals, one in eight birds, one-third of all amphibians and 70 per cent of the world's assessed plants on the List were in jeopardy. The significance of this global environmental problem is best reflected in the following statement made by Julia Marton-Lefèvre (2007), the Director General of the IUCN:

> "This year's IUCN Red List shows that the invaluable efforts made so far to protect species are not enough. The rate of biodiversity loss is increasing and we need to act now to significantly reduce it and stave off this global extinction crisis. This can be done, but only with a concerted effort by all levels of society."

Conflicts over threats to biodiversity from a range of activities associated with the human use of natural ecosystems, particularly those activities which lead to habitat degradation and fragmentation, require solutions that have their foundations in both science and law. It is for science to identify habitat that is ecologically critical for a species and whether a specific human management practice is consistent with the needs for the conservation of threatened animal and plant species. It is for law to ensure the appropriate degree of regulatory control, based on relevant and reliable evidence. In addition, the task of finding solutions for biodiversity conflicts, today, must be addressed by new alliances across all sectors of society that share an interest in conservation action for biodiversity — and not by environmentalists alone (Steiner, 2006). As Steiner points out, in order to succeed on a global scale, the responsibility for addressing the ongoing decline in biodiversity must become the responsibility of everyone with the power and resources to act.

BIODIVERSITY CONSERVATION AND SUSTAINABLE DEVELOPMENT

Science has long regarded the conservation of biodiversity and the maintenance of ecological integrity as fundamental to an ecologically sustainable society (Hare, 1991). Tarlock (2002) considers that the Biodiversity Convention may be framed as a logical extension of sustainable development, so that using biodiversity without depleting it can be seen as a strategy to achieve sustainable development. Article 10 of the Biodiversity Convention provides a focus for the requirements of sustainability and the conservation of biodiversity to be linked. For example:

i. A requirement to integrate conservation and sustainable use of biological resources into national decision-making;

ii. The adoption of measures which avoid or minimise adverse impacts on biodiversity associated with the use of biological resources; and
iii. Encouraging cooperation between governmental authorities and the private sector to develop methods for sustainable use of biological resources.

However, when conflicts over biodiversity arise, the issue of sustainability is not necessarily restricted to a strict ecocentric meaning — where only ecological considerations are to be considered. The endangered and threatened species of any nation are found on public lands ("Crown leases") and privately owned lands ("freehold") used for various human use activities such as agriculture, forestry, grazing, mining and urban development. Therefore, finding solutions for biodiversity land use conflicts through conservation measures aimed at the conservation of ecosystems upon which endangered species and threatened species depend, will invariably involve a balancing of ecological, economic, social and cultural considerations.

LEGISLATIVE APPROACHES FOR THE REGULATORY CONTROL OF THREATENED SPECIES

Analysis of approaches to defining and implementing environmental regulatory control for the conservation of biodiversity and threatened species in Australia, the United Kingdom and the USA reveals that there is common ground in the legislative schemes. All three schemes have their foundation in the following three key elements:

i. The listing of a threatened species, in different conservation categories, based on the risk of extinction;
ii. The designation of habitat critical for the survival of a listed species; and
iii. The development of a recovery plan. In the UK, this element is referred to as the "maintenance" or "restoration" of natural habitat for a species.

However, there is a major difference that exists in this sequence: in the United Kingdom and the United States listing of a species and the designation of critical habitat are both determined concurrently. For Australia, the designation of critical habitat is determined at the time a recovery plan is prepared and so is out of phase with the listing process. A further distinction that can be made is the extent that sustainable development is a matter that must be considered in the administrative decision-making process for biodiversity conservation, as either a legal obligation under national endangered species legislation or as a policy requirement.

(a) Australia

The key Federal statute is the *Environmental Protection and Biodiversity Conservation Act 1999* (the "EPBC Act"). One of its statutory objects is "*to promote the conservation of biodiversity*" [s. 3(1)(c)].

Categories of threatened species
The EPBC Act has the following categories of threatened species (at Section 178) for which a species may be included:

- Extinct;
- Extinct in the wild;
- Critically endangered ("*extremely high risk of extinction in the wild in the immediate future*");
- Endangered ("*very high rate of risk of extinction in the near future*");
- Vulnerable ("*a high risk of extinction in the wild in the medium-term future*");
- Conservation-dependent (in the absence of a specific conservation programme the species would become vulnerable, endangered or critically endangered within a five-year period).

At the end of June 2005, there were 382 listed species of fauna under the various categories for threatened species — including 14 critically endangered species, 121 endangered species and 192 vulnerable species (Department of the Environment and Heritage, 2005). In addition to the list of threatened species, a list of key threatening processes may be established. A process is threatening "*if it threatens or may threaten the survival, abundance and evolutionary development of a native species*" [ss. 183(1), 188(3)].

Critical habitat
Critical habitat is defined under the EPBC Act to mean habitat that is critical to the survival of a listed threatened species (s. 207A). Critical habitat for a listed threatened species is not identified at the time of listing — but at the time a recovery plan is made.

Recovery plans and the administrative decision-making process
The purpose of a recovery plan is to protect, conserve and manage a listed threatened species [s. 269A(2)] according to procedures prescribed in the EPBC Act. Not all species listed under the EPBC Act have a recovery plan

made, so that critical habitat will not have been designated. The EPBC Act provides the following prescribed time lines for recovery plans to be prepared after the listing of a species: within two years of a species being listed as critically endangered, within three years of being listed as endangered and within five years of being listed as vulnerable (s. 273). However, the preparation of recovery plans, and hence the designation of critical habitat, is not fully integrated with the listing of species. For example, at the end of June 2005, only 51% of the total number of listed species and ecological communities had been covered by recovery plans or had plans that were in preparation (Department of the Environment and Heritage, 2005).

The content of a recovery plan, as prescribed in Section 270(2), requires ten matters that all "*must*" be addressed in the administrative decision-making process, i.e. they are all mandatory considerations. For example: "[to] *identify the habitats that are critical to the survival of the species or community concerned and the actions needed to protect those habitats*" [para. (d)]. In addition, the making of a recovery plan also requires "*regard must be had to*" five additional matters [at Section 270(3)]. These matters for which "*regard must be had to*" include:

i. Minimising any significantly adverse social and economic impacts consistent with the principles of ecologically sustainable development [para. (c)];
ii. Meeting Australia's international agreements relevant to the species [para. (d)]; and
iii. The interests of Indigenous people in the conservation of biodiversity [para. (e)].

The phrase *"have regard to"* has been considered judicially by Australian courts on many occasions and it requires the administrative decision-maker to take those matters into account and give weight to them as a fundamental element in making his or her recommendation: per Gibbs CJ in *R v Toohey & Anor ex parte Meneling Station Pty Ltd & Ors* (1982) 158 CLR 327. Therefore, the administrative decision-maker will have discretion under Section 270(3) of the EPBC Act as to what weight will be attached to each of the five prescribed matters after they have been considered.

Accordingly, the making of a recovery plan under the EPBC Act prescribes some matters that are mandatory, e.g. the designation of critical habitat — whereas other prescribed matters are discretionary, e.g. social and economic impacts. From the perspective of the administrative decision-maker, a combination of ten mandatory matters and five discretionary matters that need to be considered for the making of a recovery plan — either singly or in combination — may well lead to a major dilemma. Not only is the integrity of the decision-making process made more complex, but also great care and rigour will be required to prepare a recovery plan to ensure that it has been prepared consistently with the principles of administrative law in order to avoid the potential threat of a judicial review challenge.

The EPBC Act and the Biodiversity Convention

With respect to threatened species, the EPBC Act expressly refers to the following aspects of the Biodiversity Convention:

i. A declaration relating to listed threatened species that will promote the survival and/or enhancement of the conservation status of the species, as well as the accreditation of a management plan made for the purpose of such a declaration, must be consistent with Australia's obligations under the Biodiversity Convention [s. 34D(1)(2)];
ii. A bilateral agreement containing a provision relating to a listed threatened species that will promote the survival and/or enhancement of the conservation status of the species must be consistent with Australia's obligations under the Biodiversity Convention (s. 53);
iii. Decision-making involving whether or not to approve the taking of an action that "*has, will have or is likely to have a significant impact*" on a listed threatened species must be consistent with Australia's obligations under the Biodiversity Convention (ss. 18, 139); and
iv. Regulations may be made in relation to giving effect to the Biodiversity Convention [s. 520(3)(i)].

(b) The United Kingdom

Bell and McGillivray (2006) state that the European Union's "Habitats Directive" (Directive 92/43/EC *on the conservation of natural habitats and of wild fauna and flora*) has had a direct effect on habitat conservation law in the UK and that it will continue to be of major significance over time. The enactment of the *Conservation (Natural Habitats, etc.) Regulations 1994* (the "Natural Habitats Regulations") implemented the Habitat Directive as domestic legislation in the UK.

The aim of the Habitats Directive is to ensure biodiversity through the conservation of natural habitats of wild fauna and flora by measures to maintain or to restore a favourable conservation status for natural habitats. Such measures are to take into account economic, social and cultural conditions.

Key elements of the Habitats Directive

- *Categories of conservation status*: "*Species of Community interest*" means species that have been placed in categories according to the risk of existing populations, in order of increasing risk, varying from "*endemic*" to "*rare*" to "*vulnerable*" and, finally, "*endangered*" [Article 1(g)].
- A "*priority species*" is also recognised as a species within the "endangered" category but, in view of the proportion of its natural range which falls within the European territory, the Community has particular responsibility for its conservation [Article 1(h)];
- *Criteria for evaluating conservation status*: Under Article 1(i), the "*conservation status of a species*" is taken as "*favourable*" by a consideration of:
  i. Population dynamics and long-term viability;
  ii. The extent of the species' natural range; and
  iii. A sufficiently large habitat to maintain a species population over the long term.
- *Listing of sites by Member States*: According to the criteria in the Habitats Directive (Annex III, Stage 1) and relevant scientific information, each Member State proposes a list of sites indicating which natural habitat types (in Annex I) and which species (in Annex II) that are native to its territory the sites host [Article 4(1)];

  In *R v Secretary of State for the Environment, Transport and the Regions, ex parte First Corporate Shipping Ltd (Worldwide Fund for Nature UK and Avon Wildlife Fund), intervening* [2000] All ER (D) 1712, the European Court of Justice concluded that:

  > "[Under] a proper construction of Article 4(1) of the Habitats Directive, a Member State may not take account of economic, social and cultural requirements or regional and local characteristics as mentioned under Article 2(3) when selecting and defining boundaries as proposed to the Commission as eligible for identification as sites of Community importance."

- *Preparation of draft list of sites by the Commission*: A draft list of sites of Community importance is drawn from the lists of the Member States to indicate those sites where one or more priority habitat areas, or priority species, have been lost [Article 4(2)]. A site of Community importance is one in the biogeographical region that contributes significantly to the maintenance or restoration, at a favourable conservation status, of the natural habitat type or of a species [Article 1(k)];

- *Designation of priorities for maintenance or restoration of habitat or a species:* Following the adoption of the list of sites of Community importance by the Commission [Article 4(2)], priorities for the maintenance or restoration, at a favourable conservation status is to be designated according to the threats of degradation or destruction to which the sites are exposed [Article 4(4)];
- *Designation of special areas of conservation:* Where a site of Community importance has been designated by Member States as representing a "*special area of conservation*", Member States through a statutory, administrative and/or contractual act are to apply the necessary conservation measures to maintain or to restore the natural habitats and/or species at a favourable conservation status. Member States are required to establish specific management or development plans for the sites that correspond to the ecological requirements for the habitat and species present on the site [Article 1(l) and Article 6(1)];
- Where there is a negative assessment of the implications for a site from a plan or project and where there are no alternative solutions — but nevertheless the plan or project must be carried out for imperative reasons of overriding public interest, including social and economic considerations, Member States are required to take all compensatory measures necessary to avoid the deterioration of habitats and species, or disturbance to species, in special areas of conservation [Article 6(4)];
- *Omissions of priority habitats or priority species in the National List:* An "*exceptional case*" is recognised [Article 5(1)] where the National List fails to mention a priority habitat or priority species. A bilateral consultation process is then initiated between the Member State and the Commission in which all scientific data are compared, in order to resolve the issue [Article 5(1)];
- *Obligations on Member States:* Member States are required to take the necessary measures to establish a system of strict protection for listed animal species in Annex IV(c) (Article 12) and for listed plant species in Annex IV(b) (Article 13).

The Habitats Directive and the Biodiversity Convention

The Habitats Directive recognises a number of the principles in the Biodiversity Convention that relate to biodiversity conservation. Specifically: (i) the need to protect both threatened species and their habitats; (ii) the maintenance and recovery of viable populations of species in their natural surroundings; and (iii) the adoption of measures to minimise adverse impacts on biodiversity associated with the use of natural resources.

The Habitats Directive and national law

The Habitats Directive is considered to provide a more rigorous legislative approach for conserving habitats and sites compared with the previously existing national law (Bell and McGillivray, 2006). The Natural Habitats Regulations apply to "*European Sites*" according to its meaning as defined in Reg. 10, e.g. "a special area of conservation". Part II of the Regulations (Regs 7 to 36) provides for "*The Conservation of Natural Habitats and Habitats of Species*", specifically, "European Sites", "Register of European Sites", "Management Agreements", "Control of Potentially Dangerous Operations" and "Special Nature Conservation Orders".

Whilst there is considerable overlap between the Natural Habitat Regulations and the *Wildlife and Countryside Act 1981*, Stookes (2005) draws the following distinction. The Natural Habitat Regulations, in implementing the Habitats Directive, result in improved protection for animals as the scope of the legislation includes deliberate disturbance. In contrast, in the *Wildlife and Countryside Act*, disturbance relates to animals when under shelter.

(c) The United States

A key purpose of the *Endangered Species Act of 1973* (the "ESA"), 16 U.S.C.A (Sections 1531 to 1544), is "*to provide a means whereby the ecosystems upon which endangered and threatened species depend may be conserved* [and] *to provide a program for the conservation of such endangered species and threatened species*" [Section 1531(b)]. There are a number of key interrelated elements in the statute that are directed at achieving this purpose:

- The listing of species in two categories — "*endangered*" or "*threatened*";
- A requirement for the designation of the critical habitat of the species to be made at the same time as the listing of the species [Section 1533(3)(A)]; and
- Statutory duties aimed at protecting the listed species in order to enable their continued existence.

Listing of species and conservation categories

The decision to list a species is based on scientific facts alone and is quite independent of any economic impacts associated with listing. The ESA (at Section 3) provides the following definitions for the two categories of listed species:

- Endangered species: "*any species which is in danger of extinction throughout all or a significant portion of its range …* " [Section 1532(6)]; and
- Threatened species: "*any species which is likely to become an endangered species within the foreseeable future throughout all or a significant part of its range*" [Section 1532(20)].

Critical habitat

The ESA [Section 1532(5)] provides for a critical habitat to be designated for species listed as threatened or endangered — but for which no critical habitat has been established. As defined in Section 1352(5)(C), the meaning for critical habitat does not necessarily include the entire geographical area occupied by the listed species. The designation of any habitat considered to be critical habitat for the listed species is made at the same time as the determination to list a species [Section 1532(5)(A)] — notwithstanding the designation may be revised where appropriate [Section 1532(5)(B)]. The ESA requires critical habitat to be designated at the time of listing in order to prevent a "taking", i.e. to harm, hunt, kill, capture, trap [Section 1532(19)] of the listed species in the designated area (Ferrey, 2004).

In contrast to the requirements for listing a species to be based on scientific facts only, the designation of critical habitat is based on a consideration of the "*best scientific data available*" and the "*economic impacts*" of specifying an area as critical habitat: Section 1533(2). However, the paramount consideration under the ESA is to avoid any failure to designate an area as critical habitat that will result in the extinction of the species concerned. In this regard, a somewhat controversial issue relates to any administrative decision made to exclude an area of habitat from protection. An administrative decision to exclude habitat requires a consideration and balancing of the best available scientific data and any other relevant impacts, e.g. economic impacts. An area may be excluded from critical habitat if the benefits of such inclusion outweigh the benefits of specifying the area as part of the critical habitat: Section 1533(2).

Statutory duties: recovery plans

Recovery plans represent one statutory duty to protect listed species under the statute. The goal of a recovery plan [Section 1533(f)] is the conservation and survival of a listed species. In order to develop and to implement a recovery plan, Section 1533(f) requires the following elements to be addressed to "*the maximum extent practicable*":

- Priority given to endangered species that are most likely to benefit from a recovery plan;

- Each recovery plan is required to incorporate a description of site-specific management actions necessary to achieve the statutory goal of a recovery plan;
- The time and cost taken to achieve the statutory goal of the recovery plan are required to be estimated; and
- Objective, measurable criteria are required to be contained in the recovery plan, in order to determine when a species may be removed from the list.

Apart from the statutory duty prescribed under the ESA to develop a recovery plan there is also a duty imposed on a Federal agency to ensure that any agency action is not likely to jeopardise the continued existence of a listed species or to result in the destruction or adverse modification of the habitat of such species [Section 1536(2)]. The statutory duty under Section 1536(2) is not imposed on private landholders.

Concerns associated with the Endangered Species Act

Notwithstanding that the ESA has been in existence for over 30 years, a number of significant concerns continue to be raised in relation to critical habitat under the ESA:

- The protection of a listed species only commences after it is listed. However, there is difficulty in achieving the purposes of the ESA because the "*habitat listing provisions are vague and often counter-productive*" (Tarlock, 2002); and
- The fact that many listed species still do not have a critical habitat designated is a continuing problem, and contrary to the authority of the ESA (Hardy, 1992; Ferrey, 2004).

Recent proposed amendments to the ESA

The ESA remains a contentious statute, especially in terms of its effectiveness in recovering endangered species as well as its impacts on landholders. On 19 September 2005 a Bill was introduced into Congress (H.R.3824: *Threatened and Endangered Species Recovery Act of 2005*) that sought, amongst other things, to eliminate the critical habitat requirement of the ESA — but provided no alternative to ensure the protection of habitat needed for species recovery (Union of Concerned Scientists, 2006). However, the Bill never became law (GovTrack.us, 2007).

On 16 February 2007, a Bill was introduced in the 110th Congress — *S658: Endangered Species Reform Act of 2007* (Senator C. Thomas), aimed at amending provisions of the ESA in relation to listing, recovery planning and delisting of species (Red Lodge Clearing House, 2007).

(d) Biodiversity Conservation, Sustainable Development and Administrative Decision-Making

It is particularly significant that the national legislative scheme in Australia incorporates biodiversity conservation within the concept of sustainable development. Section 3A of the EPBC Act prescribes a number of principles of ecologically sustainable development including that "*the conservation of biological diversity should be a fundamental consideration in decision-making*" (at para. d). For the UK, sustainable development is prescribed as a task of the Member States (as Article 2) of the consolidated version of the *EC Treaty Establishing the European Community (2004)*. In addition, Article 6 gives effect to sustainable development in terms of Community policies and activities. Also, the national environment policy for the UK, *Securing the Future — Delivering UK Sustainable Development Strategy* (2005), incorporates biodiversity conservation within the concept of sustainable development as Guiding Principle 1 ("Living Within Environmental Limits"): "*Respecting the limits of the planet's environment, resources and biodiversity*".

It is quite clear that for Australia and the UK, at least, there are statutory obligations and policy needs for the conservation of biodiversity to be considered in relation to sustainable development. These statutory and policy requirements place a heightened requirement on the integrity and consistency of the administrative decision-making process by Government agencies where decisions and actions relating to critical habitat and recovery plans for listed species are made. There is no Federal legislation in the United States that prescribes sustainable development as a matter for which there is a legal obligation for it to be considered in relation to the regulatory control of biodiversity.

INTERNATIONAL ENVIRONMENTAL CONVENTIONS AND BIODIVERSITY

The Biodiversity Convention involves almost universal global agreement as it involved 190 of the Member States of the United Nations. In terms of the Biodiversity Convention becoming legally binding on them, as at 7 May 2007, 187 nations had either "ratified" (158 nations) "approved" or "accepted" (8 nations) or "acceded to" (21 nations) the Convention (Secretariat of the Convention on Biological Diversity, 2007b). Whilst Australia, the United Kingdom and the United States were all signatories to the Biodiversity Convention, only the parliaments of Australia and the United Kingdom have ratified it. The United States remains one of the three global signatories yet to ratify the Biodiversity Convention as the Senate has yet to ratify it. The Senate process ultimately stalled in the final years of the Clinton administration and it appears that there are no current efforts to enable its ratification (Brunnée, 2004). Consequently, the United States is not formally bound by the Biodiversity Convention. The situation for the United States is succinctly described by Tarlock (2002). That is, the United States needs to ratify the Biodiversity Convention in order to establish biodiversity conservation as an overarching legal objective. In turn, the development of a comprehensive national biodiversity conservation strategy would be facilitated.

The objectives of the Biodiversity Convention (Article 1) extend beyond the conservation of biodiversity through its recognition that ecological interrelations between species are essential for their survival, i.e. the need to protect both threatened species and their habitats. Additional objectives focus on sustainable use for the categories of biodiversity components, e.g. ecosystems and habitats, species and communities and described genes and genomes, as specified in Annex I of the Biodiversity Convention, as well as the fair and equitable sharing of the benefits arising out of the utilisation of genetic resources. In addition, Article 10 also recognises the needs for Indigenous peoples and the local community by requiring each signatory nation:

i. To protect and encourage the customary use of biological resources in accordance with traditional cultural practices that are compatible with the requirements for conservation or sustainable use (para. c); and
ii. To support local populations to develop and implement remedial action in degraded areas where biological diversity has been reduced.

THREATENED SPECIES: LEGAL RIGHTS AS A SOURCE OF KNOWLEDGE POWER FOR NEGOTIATING SOLUTIONS

Environmental regulatory control for biodiversity and threatened species in Australia, the United Kingdom and the United States is heavily codified in legislation. The legislative schemes set the boundaries for the statutory powers that Government agencies must adhere to in their administrative decision-making process for actions that relate to the three key elements for the conservation of threatened species: (i) the listing of a threatened species in different conservation categories based on the risk of extinction; (ii) the designation of habitat critical for the survival of a listed species; and (iii) the development of a recovery plan or the maintenance or restoration of habitat or species. Administrative decisions made by Government agencies commonly relate to the issue of permits or licences under threatened species legislation. Also, some of the legislative approaches may impose a public duty that a Government agency must perform. Any failure by an agency to perform a statutory duty, that is mandatory, constitutes an illegal action. Consequently, civil enforcement action may be undertaken under a number of pathways to challenge administrative decisions made, or actions taken, by Government agencies under threatened species legislation.

Possible grounds for a judicial review may relate to an error of law in the administrative decision made by a Government agency. For example, whether the decision was authorised by the statute by taking into account the matters prescribed for the meaning of "threatened species", "critical habitat" and "recovery plan", as defined in the relevant legislation — or whether procedures that were required by law to be observed in connection with the making of a decision were observed. Also, where there is a legal obligation for the Government agency to perform a public duty prescribed by the statute for a decision made to grant a licence or permit, such a decision warrants close review.

The scientific literature dealing with habitat conservation is generally recognised as a scientifically accepted body of knowledge because of widespread acceptance of the scientific principles for habitat and threatened species conservation. However, the problem is that information conflicts will exist because of lack of information, different interpretations of the same information and scientific uncertainty. In this situation, grounds for judicial review in Australia and the UK could consider whether there was no evidence to support the findings of fact, or an error of material fact, in the administrative decision made. In the United States, the grounds for review, in this situation, may be whether the administrative decision could be substantiated in relation to the "arbitrary, capricious or abuse of discretion" test as well as the requirements for substantial evidence and the "hard look" doctrine. It is quite clear that great care and rigour will be required by a Government agency to avoid these grounds for a judicial review and will require the agency to carefully evaluate the scientific information against all the matters that are prescribed as matters to be taken into account as a legal obligation in the threatened species legislation.

Finally, threatened species legislation represents a very significant source of power for parties when the threat of a judicial review challenge is used as a trigger to initiate the use of ADR and negotiation to resolve conflicts over the conservation of threatened species. For parties relying on knowledge of legal rights as a source of power, it provides them with an adequate basis of power to participate effectively in resolving the conflict by negotiation.

Moreover, the situation for Australia and the United Kingdom, compared with the United States, is that knowledge of legal rights is even more significant as a source of power as legal rights for threatened species extend, to include:

- The principles for biodiversity conservation contained in the Biodiversity Convention;
- Domestic legislation implemented to give effect to the Biodiversity Convention; and
- Biodiversity conservation being linked to sustainable development.

In these circumstances, the onus on the administrative decision-maker to avoid a judicial review challenge becomes even more rigorous, as any decision made for threatened species must also be necessarily consistent with international obligations for biodiversity conservation as well as sustainability.

SCIENCE, LAND DEGRADATION AND HABITAT CHANGE

Science has long been aware that the primary cause for the loss of biodiversity is habitat fragmentation and land degradation (Shaffer, 1987). Disturbance to natural ecosystems that lead to land degradation can arise from human use activities, e.g. agriculture, forestry and mining, as well as from natural causes, e.g. drought, fire, insects and pathogens. The impacts on habitat and biodiversity following disturbance are dependent on the time the disturbance persists, the intensity of the disturbance and the spatial extent of the disturbance. Impacts may arise as an "acute disturbance", i.e. a catastrophic event over a short time-scale, such as clear felling of a natural forest. Alternatively, impacts may arise as a "pulse disturbance", i.e. low intensity or intermittent impacts over a long time-scale, such as selective logging over large areas of land. Pulse disturbance events are much more difficult to study experimentally, relative to acute disturbance events, and so pose a special problem for scientific proof of causation.

The processes of land degradation have been described by Christie (1993e) as involving a continuum of change from slight to severe deterioration of the biological and soil resource. The early warning signs of land degradation are primarily changes in species diversity of the plant community (Figure 9.1). It is at this stage that plant species most sensitive to disturbance, or indicator species, are affected. As the disturbance continues, other plant species are lost, so that the impacts result in a loss in both plant species diversity and ground cover. Wildlife populations are affected at the stage when their habitats commence to degrade, or fragment, because of the disturbance. Depending on the time of and the extent of disturbance, fragmentation of habitats can lead to the loss of species from habitat remnants, e.g. when the remnants are too small to support a viable population or when the remnant supports a plant population but not its pollinator.

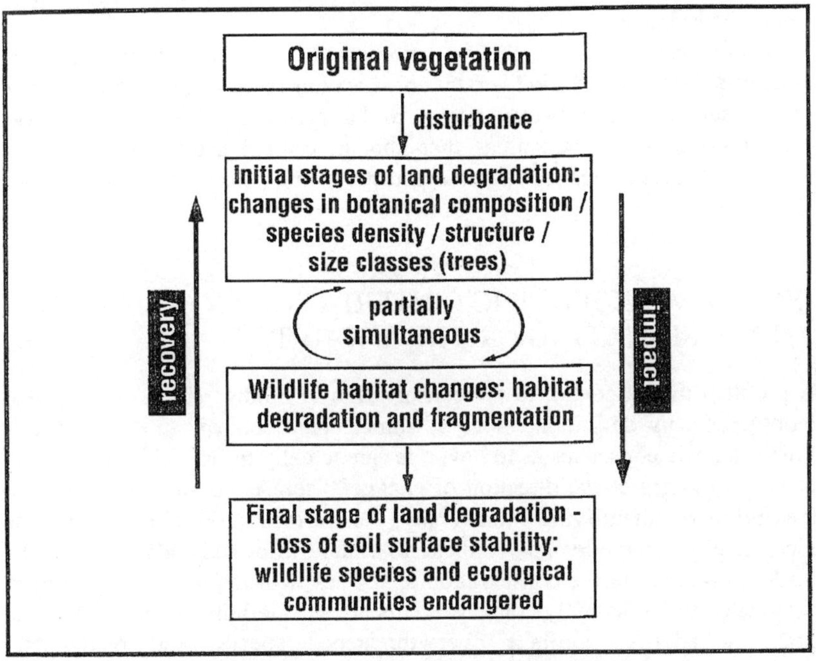

Figure 9.1 Generalised relationships between the processes of land degradation and wildlife conservation (Christie, 1993e)

Should the disturbance continue to persist over time, the end-point of land degradation will eventually be reached; i.e. the loss of most, if not all the plant cover and loss of the stability of the soil surface. At this stage, regeneration of plants becomes a very slow, difficult process; the central question then becomes whether the natural ecosystem will return to, or near to, its pre-existing biological state prior to the disturbance. If this is not the case then a species or ecological community would become threatened.

Resilience is an accepted ecological concept that is central to understanding whether ecological impacts, following disturbance, are either reversible or irreversible following the cessation of the disturbance. Resilience is dependent on (i) the pattern of response by an ecosystem to disturbance, as well as (ii) the capacity of the ecosystem to recover when the disturbance is removed. As all natural ecosystems vary in their "resilience", it is a key factor in the consideration of the designation of critical habitat of a listed species. In addition, the measures required to avoid or minimise adverse impacts on biodiversity associated with the use of biological resources. Resilience, together with the concepts of sustainability and "ecological health", are the foundations for the scientific management of natural ecosystems.

From an ecological perspective, incorporating sustainability in relation to threatened species and biodiversity conservation requires the habitat critical to the survival of a listed species to be preserved. Furthermore, large contiguous or connected areas that contain critical habitat must also be retained. Opportunities for spatial connectivity must also be promoted (Dale *et al.*, 2000).

OBJECTIVE SCIENTIFIC CRITERIA FOR EVALUATING AND DESIGNATING CRITICAL HABITAT

A problem that is central to the conservation of biodiversity is the fact that biodiversity, by definition, is an 'n'-dimensional concept and that there is little scientific consensus as to how one can actually measure biodiversity — or how to interpret the direction of change. There is no simple method for measuring or quantifying biodiversity. As Tarlock (2002) points out, the identification of biodiversity indices to enable informed judgements to be made, in relation to the conservation of species, will require "*heroic efforts*" by science as biodiversity conservation is not self-defining. Consequently, in order to address conflicts over threatened species and biodiversity conservation, the paramount need is to understand and to apply the accepted body of scientific knowledge for habitat conservation and the measures that need to be adopted to avoid or minimise adverse impacts on biodiversity associated with the use of biological resources.

For example, science has long been aware that a basic need for threatened species and biodiversity conservation is to identify, then protect, habitat that is critical for the survival of a listed species; in turn, to restrict any consideration of development proposals to the least sensitive habitat (Snape, 1996). In this regard, some of the key factors that represent an accepted body of scientific knowledge for defining critical habitat have been described by Olson (1996). The evaluation of critical habitat needs to focus on two factors: (i) habitat quality and (ii) conservation value. The element of habitat quality is equally important as conservation value for any assessment of the habitat that is critical for the survival of a listed species:

- The term "habitat quality", "reflects the extent to which the physical, biological and other environmental characteristics of an area correspond to the habitat characteristics of the [listed] species" (Olson, 1969).
- The term "conservation value" means "the ability of an area to support a [listed] species over the long term" (Olson, 1969).

(a) Habitat Quality

For an individual listed species, habitat quality may be considered in terms of scientific data in relation to:

- *Physical characteristics* — such as soil type, slope, aspect and elevation; and
- *Biological characteristics* — such as vegetation structure (trees, shrubs, herbs), species composition and diversity or the presence of "indicator species" that reflect the biological condition (or status) of the area in response to disturbance, e.g. land degradation.

The process of evaluation of habitat quality leads to areas being identified that contain a unique set of physical and biological characteristics necessary to support a listed species. However, the requirements for the conservation of biodiversity and threatened species are also dependent on complementing scientific data on habitat quality with data on conservation value.

(b) Conservation Value

Dale et al. (2000) and Olson (1996) provide the following objective criteria for evaluating conservation value:

i. The size of the area;
ii. Configuration (or shape of the area), e.g. a "rounder" shape is preferred to a shape that is linear or irregular;
iii. Spatial connectivity of patches of habitat that have become fragmented following human use activities, e.g. land-clearing for agriculture or human settlements — or through natural causes such as fire, drought or non-native species;
iv. Length and width of habitat corridors;
v. Exposure of habitat to unprotected edges;
vi. Threats from non-native species; and
vii. The role of unoccupied habitats and habitat refugia in population persistence.

RESOLVING BIODIVERSITY AND THREATENED SPECIES CONFLICTS: ALTERNATIVE APPROACHES

Threatened species occur on land under private ownership (or freehold land) as well as on public lands (or Crown land leases), used for a wide range of human use activities, e.g. agriculture, grazing, timber, mining, infrastructure for cities and towns, and recreation. It is inevitable that solutions for conflicts over the conservation of threatened species and biodiversity will frequently involve ecological considerations (related to the listed species) as well as economic considerations (related to the interests of landholders).

The legislative schemes for threatened species in Australia, the United Kingdom and the United States all have provisions for economic considerations in relation to the designation of critical habitat for a listed species. Consequently, ecological and economic considerations are central to finding a solution for conflicts over the designation of habitat that is critical for the survival of a listed species. In addition, in Australia and the UK, biodiversity conservation extends to incorporate the concept of sustainable development so that ecological, economic and social considerations all become matters required to be taken into account in the administrative decision-making process, as a legal obligation, for the designation of critical habitat for a listed species; in Australia, under the EPBC Act, these matters extend to Traditional Indigenous knowledge.

(a) Conflict Resolution: Scientific Approaches Focusing Primarily on Ecological Considerations

A number of innovative approaches have been used in which the focus is on habitat conservation and the need to protect habitat that is ecologically critical for the survival of a listed species.

i. "*Buyer diversity*" schemes involve the formation of Private Nature Conservation Groups and Trusts that accept donations of land, or alternatively purchase land — land that has high conservation value (Cowell and Williams, 2006). The land then becomes subject to permanent reservation. Another approach involves developing agreements with landholders to secure private land using various forms of conservation covenants. Land so acquired may be sold on to a new owner with the conservation values secured by an in perpetuity covenant on title; and
ii. The *safe harbour agreement* is a model that emerged from the ESA of the United States. The safe harbour agreement is a voluntary agreement between the Federal agency and a landowner that sets out specific

management activities that the landowner will undertake, or forego, in order to provide a "net conservation benefit" to species covered by the agreement; i.e. a benefit that contributes directly, or indirectly, to the recovery of the threatened species. In addition, the agreement provides the landowner with safe harbour assurances that no additional regulatory restrictions will be imposed on land use, in the future, because of the landowner's conservation activities. Should ownership of the land change in time, the agreement remains — provided the new landowner willingly signs the original agreement or a new mutually agreeable one (US Fish and Wildlife Service, 2004).

(b) Conflict Resolution: Scientific Approaches Focusing on Ecological and Economic Considerations

There are also a number of approaches that address both ecological and economic considerations in the context of sustainability. A special focus is on the need to enhance, restore or re-create habitat that has been fragmented or destroyed. Tradeable units of "conservation credits" to facilitate economic development at the regional level have also been proposed where the conservation of biodiversity is in issue.

EU policy: management of Natura 2000 Sites[5] for conservation

The European Community policy, titled *Managing Natura 2000 Sites: The Provisions of Article 6 of the Habitats Directive*, facilitate the interpretation of Article 6 ("*Conservation Measures and Management Plans*") of the Habitats Directive. Article 6 plays a crucial role in the management of Natura 2000 sites. As an EU policy, the guidelines are non-binding. However, their scope for finding solutions for threatened species conflicts is that they provide a detailed framework for (i) finding creative solutions for mutual gain through interest-based negotiation and (ii) achieving the purpose of threatened species legislation and sustainability. The policy guidelines contain, amongst other matters, "*compensatory measures*"; compensatory measures do not mean direct economic measures, but rather achieve this aim indirectly. The aim of a compensatory measure is to compensate for any negative impacts on habitat arising from a development project or plan and include:

- Re-creating a habitat on a new or enlarged site to be incorporated into Natura 2000;

[5] The establishment and conservation of the EU network of sites designated by Member States under the Habitats Directive and the "Birds Directive" (Directive 79/409/EEC *on the conservation of wild birds*) is known as Natura 2000.

- Improving a habitat on part of the site, or in another Natura 2000 site, proportional to the loss due to the project; and
- In exceptional cases, proposing a new site under the Habitats Directive.

These measures are subject to a temporal (time) qualification. The outcome of the measure has normally to be operational at the time when the damage affects the project site, i.e. unless it can be proved that a simultaneous connection between the operation of the measure and negative habitat damage being offset at the project site would not be necessary to ensure the contribution of the site to the Natura 2000 network.

Compensatory measures are seen as the last resort and are used only when other safeguards provided for by the Habitats Directive are ineffective. For example, alternative solutions and mitigation measures that will result in the plan or project adversely affecting the integrity of the site. In addition, where a decision has been made to consider a project or management plan that has a negative effect on the Natura 2000 site.

The Habitat Transaction Method

The Habitat Transaction Method (Olson *et al.*, 1993; Olson, 1996) is a creative approach for incorporating ecological and economic considerations with the needs for the conservation of biodiversity at the regional level. The Habitat Transaction Method is a variant of Wetland Mitigation Banks which emerged in the United States as a means to mitigate the loss of wetlands through the use of fill or other unavoidable impacts. The method involves tradeable units of "conservation credit".

The Habitat Transaction Method recognises the biological requirements of a species listed for protection and then creates a reserve system through habitat transactions designed to meet relevant biological objectives. The methodology of Olson *et al.* (1993) and Olson (1996) is to assess habitat quality and then to adjust it, according to conservation value criteria such as size, configuration and spatial connectivity of the site. For example, in relation to spatial connectivity, fragmented habitat is devalued relative to connected habitat. The rationale for the methodology is that objective measures of conservation value can be defined in terms of "conservation units" that represent indicators of long-term conservation value for specified conservation objectives. The key features of the method can be summarised as follows:

- Critical habitat is expressed in terms of a total amount of conservation value, i.e. in units based on the indicators of habitat quality and conservation value;

- A landholder who agrees to conserve or restore habitat within the planning area, e.g. regional scale, receives credit based on the conservation value contributed to the reserve system;
- Any landholder proposing a project that would cause a loss of conservation value must offer a number of credits based on the loss of conservation value units arising from the development; and
- Landholders who receive credits for conservation actions may use the credits to mitigate projects elsewhere within the planning area or may sell the credits to any other landholder who needs credits to mitigate project impacts.

The effectiveness of this method for resolving biodiversity conflicts will be dependent on the adequacy of the existing scientific database for the listed species. The method would be most appropriate when the existing scientific database contained sufficient objective facts that could be established directly — or indirectly by the scientific method of inference, to be able to characterise habitat quality and conservation value consistently with the standards and criteria of science.

The habitat conservation plan
This approach evolved in the United States under the ESA and sets out measures to maintain, enhance and protect the specific habitat for a listed species, e.g. by restoring habitat or by permanent protection of land. Whilst science has challenged the effectiveness of habitat conservation plans in the past, Tarlock (2002) argues that the contemporary debate has moved from the merits of the basic concepts for a habitat conservation plan to how the habitat conservation plan process can be improved — the equivalent approach that has guided the development of risk assessment in the United States.

(c) Conflict Resolution: Approaches to Biodiversity Conservation Focusing on Understanding LandHolder Needs

A controversial and difficult issue for any conflict resolution process is the need to balance the ecological requirements for habitat conservation for the survival of listed species with the protection of economic activities of landholders. Tarlock (2002) contends this to be a critical aspect for habitat conservation of threatened species as it may represent a situation in which landholders could conceivably control the level of biodiversity to be preserved — as well as the ecological monitoring process. In order to offset the possibility of such an outcome — given that it may be perceived as antagonistic to multi-party negotiations for biodiversity conservation, some understanding of landholder needs is extremely important; in particular, measures that could be considered as part of a negotiated agreement for the conservation of biodiversity and economic sustainability of landholders. A parliamentary inquiry in Australia into the effects upon landholders and farmers of public good conservation measures imposed by Australian Governments (House of Representatives Standing Committee on Environment and Heritage, 2001) provides some insight into this issue. A number of measures for achieving economic sustainability that landholders identified as being useful when carrying out conservation activities on farms and grazing lands, included:

- Financial assistance to motivate as well as to facilitate the ongoing management of land, e.g. stewardship payments, rate relief, fencing and labour assistance;
- Provision of transition assistance to ecologically sustainable forms of production and management;
- Compensation for decreases in land value or loss of property rights; and
- Reform of the taxation system.

In the United States, it is particularly relevant that a Bill introduced in the 110th Congress on 28 February 2007 — *S700: Endangered Species Recovery Act of 2007* (Senator M. Crapo), is aimed at amending the Internal Revenue Code to provide a tax credit to individuals who enter into conservation agreements to protect or improve the habitats of endangered and threatened species (Red Lodge Clearing House, 2007).

TRADITIONAL INDIGENOUS KNOWLEDGE AND CONFLICT RESOLUTION

Chapter 15 of Agenda 21 deals with the "Conservation of Biological Diversity". As a basis for action, Part 15.4(g) requires Governments, at the appropriate level, to take into consideration Indigenous people and their communities, as well as social and economic factors, so as to:

> "Recognize and foster the traditional methods and the knowledge of indigenous people and their communities, emphasizing the particular role of women, relevant to the conservation of biological diversity and the sustainable use of biological resources, and ensure the opportunity for the participation of those groups in the economic and commercial benefits derived from the use of such traditional methods and knowledge".

However, Agenda 21 is a non-binding text and a signatory nation must implement domestic legislation, as considered appropriate, to give effect to it. In the case of Australia, the Federal statute, the EPBC Act, addresses Part 15.4(g) of Agenda 21. Two of the statutory objects of the EPBC Act are: (i) to recognise the role of Indigenous people in the conservation of Australia's biodiversity and (ii) to promote the use of Indigenous people's knowledge of biodiversity: s. 3(1)(f)(g). In addition, there is a legal obligation in the administrative decision-making process when recovery plans are prepared, to have regard to the interests of Indigenous people in the conservation of biodiversity: s. 270(3)(e). The effect of these statutory provisions in the EPBC Act is that traditional knowledge on biodiversity conservation can be applied together with accepted knowledge from the "Western" scientific model – given that these statutory rights represent a significant source of power to facilitate participatory management between the Federal Government agency and Indigenous people.

FINDING SOLUTIONS FOR BIODIVERSITY CONFLICTS: LITIGATION V NEGOTIATION

The 2006 and 2007 IUCN Red List studies clearly illustrate that, notwithstanding the implementation of threatened species legislation, the loss of threatened species and biodiversity has continued over time and remains as a major source of global concern. The IUCN analysis highlights the need for the use of alternative processes to litigation to resolve biodiversity conflicts, if this decline is to be effectively addressed.

(a) Litigation

The statutory listing of a threatened species, together with the designation of its critical habitat, will not necessarily ensure the survival of a listed species. In addition, recovery plans as a statutory obligation for conserving threatened species will be ineffective where there has been severe fragmentation or destruction of the habitat that is critical for the survival of a listed species; the experience in the United States serves as a good illustration of the limited success arising from recovery plans (Hardy, 1992; Holst, 1992). Based on the global trends in biodiversity decline in the 2006 and 2007 IUCN Red List studies, it could be concluded that reliance on environmental regulatory control of threatened species and litigation has not been an effective means for achieving the objective of conserving listed species or the critical habitats upon which they depend. Finally, litigation is not a process to "*find ways to incorporate the value of biodiversity into local, regional, national and even international economics in order to make private property an ally in the preservation and conservation of biodiversity*" (Olson, 1996).

(b) Community Consultation and Public Participation

In Australia, in April 2006, the Federal Minister for the Environment refused to issue a permit under the EPBC Act for a $200 million major wind farm project in regional Australia. There were environmental benefits associated with the development of renewable energy. However, the Ministerial decision was based on threats posed by the coastal wind farm to the long-term survival of a listed species, the endangered Orange-bellied Parrot (*Neophema chrysogaster*). The scientific evidence in a report taken into account by the Minister included that this parrot was at risk of becoming extinct within 50 years without any impact from the wind farm. Whilst the impact of wind turbine collisions on this parrot was small (up to one bird death/year), the report concluded that any negative impact on the species could be sufficient to tip the balance against its continued existence (Department of the Environment and Heritage, 2006a).

However, eight months later, the Federal Minister for the Environment reversed the earlier decision and approved the wind farm project. The Minister concluded that there was no longer an unacceptable risk to the Orange-bellied Parrot because changes had been proposed to address the risks: including the location of the wind turbines being changed, a commitment to ecological monitoring of impacts on the Orange-bellied Parrot and other species, and the implementation of procedures to mitigate any potential impact in accordance with the latest and best advice. In making this decision, the Minister "*considered very carefully the 160-plus public comments*" received as well as scientific advice "*that the Orange-bellied Parrots could be adequately managed through the changes to, and conditions placed on, the proposal*" (Department of the Environment and Heritage, 2006b).

In the circumstances that the Ministerial decision was made, some concerns could be raised as to the effectiveness of the public participation process; specifically, whether the public had an adequate basis of power to participate effectively in this conflict. There are no prescribed procedures for public participation set out in the EPBC Act or Regulations; nor is there an express policy. Rather, reliance is placed on notice and comment provisions prescribed under the EPBC Act (Department of the Environment and Water Resources, 2007a). The reactive approach to decision-making used in this case also highlights the difficulty for ensuring a full and fair consultation with the public. The existence of detailed policy guidelines that dealt with possible measures directed at maintaining, protecting, enhancing, restoring or re-creating habitat that had been fragmented or lost, through to proposing a new site, equivalent to the EU Policy, *Management of Natura 2000 Sites for Conservation*, would have also facilitated the public to make an objective and fully informed response to the wind farm proposal in their submissions.

Consequently, it could be concluded that sole reliance on notice and comment provisions, where the legal obligation was to consult not to agree, has the potential to lead to a perception that the goal of public participation may be to manipulate rather than to involve the public in resolving conflicts over the conservation of biodiversity. Any such perception would have negative consequences for trust in the Government agency.

(c) ADR and Interest-Based Negotiation

ADR and interest-based negotiation, through its collaborative fact-finding and problem-solving approach — procedures that are not available through litigation and the courts — facilitates the evaluation and adoption of a wide range of creative solutions for mutual gain. Negotiated agreements are not restricted to measures available under the legislative scheme, but can include, for example, policy measures for the conservation of threatened species and biodiversity, e.g. the *EU Policy: Management of Natura 2000 Sites for Conservation.* In addition, in terms of sustainability, ADR and interest-based negotiation enables the evaluation of a number of options in which the balance between ecological, economic, social and cultural considerations varies.

Understanding the needs and concerns of all parties having legitimate interests in biodiversity conservation, e.g. the community, conservationists, Indigenous peoples and landholders, is the underlying basis for the use of ADR and interest-based negotiation. Creative solutions for mutual gain can then be directly evaluated against the needs and concerns of the parties by reviewing accepted knowledge on measures for the conservation of threatened species and biodiversity, based on the "Western" scientific knowledge and, where appropriate, Indigenous traditional knowledge. The process could be structured to effectively involve parties and "*new alliances across all sectors of society*" having legitimate interests in biodiversity conservation in a manner consistent with the proposition of Steiner (2006).

(d) Conclusion

ADR, litigation and public participation can all be used to address threatened species and biodiversity problems. However, the actual choice of process will be very much dependent on whether each party has an adequate basis of power to effectively participate in each process as well as the extent there is shared responsibility in the administrative decision-making process. The challenge for Government agencies, and parties holding legitimate interest in biodiversity conservation — such as community, developer, ecological, economic, Indigenous, landholder or scientific interests — is to recognise the central role of conflict assessment and a BATNA analysis as the basis for determining the most appropriate and effective process for finding solutions to address the disturbing trend in the decline of the earth's biodiversity.

10. Managing and resolving environmental conflicts by negotiation: NIMBY or NIMBI?[6]

"How can we best resolve issues of major controversy between groups holding opposing, yet sincerely held, opinions in ways that most nearly satisfy the principles of the democratic ordeal … solutions from which all parties can emerge with some sense of gain and certainly with the knowledge that their views have properly been taken into account by the ultimate decision-maker … "

William Hayden (1991), Former Governor-General of Australia

" … existing research suggests that collaborative principles are worth implementing not only because they are likely to produce better agreements, but also because they will almost certainly enhance levels of transparency, information sharing, and cooperative behaviour in the underlying process."

Camacho (2005)

"Our diverse, democratic, technologically advanced society could not function at all without translators … [Translators] possess special powers and skills that are of immense value to environmental negotiations … A translator is a person with specialised training in science and technology who has the ability to translate the terms and concepts of [science and technology] into the language and experiences of everyday life."

Gorczynski (1991)

INTRODUCTION

Sir Eric Ashby (1978) describes the development of an environmental conflict in terms of "*a chain reaction*" that occurs in a number of stages. The first stage, the *"ignition phase*", follows when the public first become aware of a proposed action or development. Public controversy or debate associated with perceptions of adverse consequences to the environment emerges during this stage. However, whilst public conscience is excited, this alone is not

[6] Acronyms for "*Not In My Back Yard*" and "*Now I Must Be Involved*", respectively.

sufficient as public opinion has to be raised to a "temperature" that stimulates some form of response. For the Government agency responsible for making the administrative decision of whether or not to approve the proposal, there may be an inbuilt inertia to stall at this stage, in the hope that public opinion is a passing scare and will go away if it is ignored.

The second stage in the chain reaction — the "*objective phase*" — is reached when it becomes clear to the Government agency that public opposition has become galvanised, as people directly affected by the issue unite and form as a group to take blocking actions. The Government agency may then take steps to defer the implementation of the administrative decision until any public opposition has lost its urgency or could otherwise be neutralised. For example, where the proposed development or activity is very controversial, some form of community consultation or public participation process may be set up. A special feature of the stage is for the agency to have the hazards for the proposal identified and examined objectively by scientific experts in terms of how genuine and serious their consequences might be for the environment or human health. Should the approval of the proposed project or activity require a licence or permit, the Government agency will also be required to seek public comment, where there is a statutory obligation imposed under environmental legislation, e.g. the EIA process or a permit for the deliberate release of GMOs into the environment.

The final stage of the chain reaction is a "*subjective phase*", reflecting a not uncommon outcome that findings on scientific facts by experts arising from the objective phase and the public perceptions of risk do not correspond. The dilemma for the Government agency for the administrative decision that must be made is how to combine findings based on the rational scientifically based analytical methodology of the objective phase with the subjectivity of public perceptions of risk in terms of values and emotions that arise from the final stage of the chain reaction. The administrative decision needs to be a solution that effectively addresses the legitimate claims of all parties having competing interests in the environmental conflict. In the worst case scenario, and after vigorously defending the reasoning for its actions, the administrative decision made by the Government agency may be challenged in court by some form of civil enforcement action. If the opponents are successful in blocking the proposal through litigation, the agency re-commences the entire administrative decision-making process.

During the first two stages of the "chain reaction", there are opportunities in an environmental conflict to manage the sources of conflict. If conflict still persists, then the need for conflict resolution processes will arise in the final stage of the "chain reaction". Some of the key factors that need to be recognised throughout the stages of conflict development are:

i. Finding a solution for a conflict over sustainability will require information to be obtained and assessed from ecological, economic, social and cultural sources. The facts will be complex and numerous. However, scientific uncertainty will almost certainly exist, so reducing the likelihood of the parties reaching complete or unanimous agreement on all facts;
ii. Finding a solution for a conflict over sustainability will require concerns for adverse consequences to the environment or human health, based on public perceptions of risk held by parties that may be subjective or value-laden, to be balanced against the objective, science-based assessment of risk; and
iii. There will be little likelihood of reaching complete agreement on the disputed facts in these circumstances for multi-party environmental conflicts. Where negotiation is used, concessions would need to be made on disputed issues by the parties in order to find a compromise solution that best meets their interests. Where litigation is used, the court will adjudicate on the disputed facts and impose its decision on the parties.

After 24 years in politics in the State of Oregon in the United States — the last 8 years as Governor — John Kitzhaber (2004) argues that there is now a need for a new approach to the procedures and processes for resolving problems associated with environmental sustainability. Kitzhaber emphasises that problems over economic activity and environmental stewardship do not arise solely because of the people engaged in the conflict; the problem arises because of limitations in existing processes, institutions and organisational structures through which people are seeking to resolve their conflicts. Moreover, the issue is not with the environmental regulatory control standards imposed by law and enforcement — but the procedures and processes that have been traditionally applied. Not only have contemporary environmental problems become far more complex, but society is now faced with natural and fiscal limits unimaginable in the past. Paradoxically, society continues to cling, obstinately, to "*a set of tools and governance structures, created in an earlier era to address a very different set of problems*" (Kitzhaber, 2004). As Kitzhaber points out, the problem is not simply the responsibility of the legislature or the courts or the administration to do something. Finding a new approach for balanced natural resource management and environmental sustainability is based on the proposition that there is a commitment to develop good science and accurate monitoring, notwithstanding that scientific uncertainty will exist. In addition, the establishment of a credible and transparent process in which the best available science can be assessed and applied is crucial.

LITIGATION V NEGOTIATION AND ENVIRONMENTAL CONFLICT RESOLUTION

There are a number of alternative processes for managing and resolving an environmental conflict. For a conflict that is subject to litigation, settlement by the parties' lawyers prior to the case being heard by a court is the first option. If this is unsuccessful, the matter may be referred to ADR, through legislated procedures for court-annexed ADR. If a negotiated outcome does not emerge, then litigation is the final option. In addition to the various processes and procedures available through the courts, environmental conflicts can still be resolved external to and independent of the courts. A general term to define these procedures is "non-court-annexed ADR".

(a) Pre-hearing Settlement

Pre-hearing settlements are usually conducted by the lawyers acting for the parties, but in the absence of the parties. That is, the parties have no direct involvement in discussions to negotiate a settlement. Their role is to agree whether or not to accept the outcome arrived at by their lawyers. There is no neutral third party to facilitate negotiations. There is no structured process of negotiation equivalent to that found in ADR. Notwithstanding that environmental conflicts may be settled in this manner, it is possible that a better agreement could be crafted (Susskind *et al.*, 2000). To support this proposition, Susskind *et al.* have identified the following factors as possible limitations to a pre-hearing settlement agreement:

- A failure to address the underlying interests that created the conflict for the parties. Many agreements may represent little more than one side "giving up";
- A settlement agreement will be dependent on how well the parties involved are able to deal with their differences through negotiation. If this proves to be limiting or ineffective, the outcome is self-evident; and
- It is impossible to know whether settlement negotiations have been undertaken on the basis of "good faith" by all parties.

(b) Litigation

The outcome of litigation is an imposed decision by the court on the parties and a winner and a loser. Legitimate compromises are often overlooked in litigation; parties engaged in litigation are often unwilling to participate in the "give-and-take" that forms the basis for reaching a mutual agreement through interest-based negotiation. Furthermore, litigation discourages open communication, the sharing of information and joint problem-solving; a consequence of litigation is that the relationship between the parties declines (Susskind *et al.*, 2000). The central issue, therefore, becomes whether the courts should be used as a remedy of last resort used when all other conflict resolution strategies have failed — or whether litigation and adjudication should be the primary pathway for settling environmental problems. Consequently, screening a case for its suitability for negotiation or litigation is a crucial issue for part of any BATNA analysis.

(c) ADR and Interest-Based Negotiation

The alternative process to litigation is ADR and interest-based negotiation. In contrast to litigation, negotiation leads to the parties finding their own solution that meets the interests of all parties and to reach agreement by consensus. ADR and interest-based negotiation incorporates a number of elements to facilitate parties reaching the goal of finding an enduring, effective agreement that all parties can live and abide with. The principled negotiation approach of Fisher *et al.* (1991) and the mutual gains approach to negotiation (Susskind *et al.*, 2000) are the foundation for achieving this goal and are reflected in the following elements for managing and resolving a conflict using ADR and interest-based negotiation:

- Joint fact-finding;
- Joint problem-solving;
- Clarifying and understanding the interests of each party;
- Creating a number of options or proposals for mutual gain; and
- The use of objective criteria in order to evaluate the options or proposals, e.g. based on standards such as scientific merit or fairness that are independent of each side's will; legitimate and practical standards.

(d) Court-Annexed ADR v Non-Court-Annexed ADR

There are two alternative pathways for resolving environmental conflicts by negotiation when knowledge power of legal rights is the catalyst for conflict resolution:

- Court-annexed ADR, where legislation mandates that ADR first be used to resolve a matter that is the subject of litigation. A court hearing will only proceed if the conflict cannot be resolved through negotiation; or
- Non-court-annexed ADR is external to, and independent of, the courts. For example, where a Government agency is willing to share power with those affected by the administrative decision to be made in order to resolve the conflict by negotiation.

Whilst both pathways involve the use of ADR processes and negotiation, they can be distinguished in terms of their characteristic features and pre-requisites for success (Susskind, 2005). Consequently, these differences must be understood in any BATNA analysis undertaken by the parties.

Court-annexed ADR — litigation

Susskind (2005) considers that court-annexed ADR provides little advantage, relative to litigation, for complex public interest issues. Representation is restricted to only those parties who can establish they satisfy the standing requirements to litigate the matter. Other parties affected by the outcome will be excluded from the action if they have no legally recognisable interest in the conflict. Susskind (2005) provides a clear analysis of further limitations in court-annexed ADR that limits its scope for resolving public interest environmental conflicts:

- Parties involved in court-annexed ADR know that they will still have their "day in court" if negotiations fail and the matter cannot be resolved through ADR. Consequently, the parties may give priority to those issues that are the basis for the court action and so inhibit creative problem-solving; the demands of litigation may interfere with opportunities to maximise mutual gains;
- In addition, parties may be wary of committing themselves to a joint fact-finding approach or to a collaborative approach to information analysis. This may be the case if there is a perception that information may be disclosed that may disadvantage their position should negotiations fail and the matter is then set down for a court to determine; and
- Negotiation cannot guarantee that the public interest will be met where the standing rules restrict the parties who may participate.

Non-court-annexed ADR — sharing power

As is the case for court-annexed ADR, participation may be restricted to those parties who have a legally recognised interest, such as satisfying the rules for standing to litigate the matter. For some environmental conflicts, this may be appropriate, e.g. an administrative decision to grant a licence or permit for a proposed activity in which the issues are few and simple and the problem site-specific.

However, a significant difference arises for public interest environmental conflicts. In particular, where public comment on a proposed action is a mandatory requirement prescribed by environmental legislation, e.g. an EIS. Public comment provisions for actions that may cause environmental harm, and which require a permit or licence, are generally not prescribed in environmental legislation. However, some exceptions occur such as in biotechnology and threatened species legislation. Consequently, where a Government agency is willing to share its power with those affected by the administrative decision to be made, it may use discretion to widen the scope of participation beyond those meeting the standing rules. Parties who made submissions in response to the public notice and comment provisions would have a relevant interest in the conflict. Characterisation of the nature of their interest would facilitate the identification of parties who may have a role in negotiating outcomes for a public interest environmental conflict.

The advantages of the non-court-annexed ADR model for resolving environmental conflicts, compared with the court-annexed ADR model, can be summarised as follows:

- Public participation would be maximised, ensuring that the administrative decision would be made with a full understanding of the issues and the consequences of the proposal and for the public interest to be met;
- In contrast to the court-annexed ADR model, the collaborative approach to joint fact-finding and problem-solving would be undertaken independently of the litigation pathway being available should negotiations fail;
- As is the case for ADR, generally, non-court-annexed ADR provides the parties with flexibility to choose ADR procedures that would be appropriate to manage and to resolve their specific conflict. Additionally, and in contrast to courts and litigation, the parties may agree on the basis for the agreed negotiated agreement, i.e. facts, law, policy, values or some combination; and
- In common with court-annexed ADR, the process of reaching a negotiated agreement in the non-court-annexed ADR model may also be rights-based, e.g. where there is no dispute over the law. Matters to be taken into account as a legal obligation, by the Government agency, in the administrative decision-making process could be considered. Assessment of the opportunity to share power between an agency and the parties to a potential civil enforcement action would be facilitated.

THE BEST ALTERNATIVE TO A NEGOTIATED AGREEMENT AND ENVIRONMENTAL CONFLICTS

There are a broad range of alternatives for resolving any environmental conflict, such as litigation, negotiation, public participation or political lobbying, that need to be considered in a BATNA analysis. As part of the process of conflict assessment, a BATNA analysis enables a party to determine whether they have an adequate basis of power to arrive at a more favourable solution by negotiation, compared with the outcomes from other alternatives, e.g. litigation, consultation or lobbying Government. A BATNA analysis is therefore a strategic means for all parties engaged in an environmental conflict to determine how appropriate alternative processes may be for resolving the conflict. In addition, by comparing the potential of negotiation relative to alternative processes for conflict resolution, a BATNA analysis also enables each party to consider what may happen to them if a negotiated agreement cannot be achieved and the parties walk away from the negotiation process. What alternatives for conflict resolution would then exist and be available for each party — and how effective would they be for achieving goals that may have been possible through negotiation?

(a) Government Agencies and BATNA Analysis

A Government agency in exercising regulatory control may see its role in deciding applications for permits or licences under environmental legislation to be confined to strict questions of law. In addition, the legislation may not mandate the use of ADR and interest-based negotiation as a requirement for the administrative decision-making process. A civil enforcement action in the form of a judicial review to challenge the legality of the administrative decision-making process for environmental issues is an ever-present potential and contemporary threat for Government agencies. The utility of a BATNA analysis is that it enables the Government agency to determine whether negotiation, or defending a potential civil enforcement action such as a judicial review, would be the preferred process for resolving an environmental conflict. In evaluating its BATNA for an environmental conflict, a Government agency would need to consider the following factors as part of its assessment:

i. The legal principles in administrative law for the grounds for a judicial review challenge over the legality of an administrative decision are well established. The interpretation of the law on the grounds for a review should generally not be in dispute;
ii. A judicial review is seen as an effective use of legal rights to provide a check and balance against the wide statutory decision-making powers held by Government agencies for many environmental matters;
iii. Where the standing requirements for a judicial review that are provided by legislation and common law are wide, potential challenges could arise from many sources — other than from the party seeking approval for an action having environmental consequences;

iv. Environmental conflicts may be problematic for the administrative decision-maker to ensure a proper, genuine and realistic consideration has been given to the matters that must be taken into account as a legal obligation under environmental legislation;
v. Problems of scientific uncertainty, lack of information and missing information, compounded by problems in having relevant factual evidence to predict future impacts, such as novel (biotechnology) or cumulative (EIS) impacts, exist. The implication for the administrative decision-making process is whether, in these circumstances, a judicial review may be successful on the grounds of the "no evidence rule/unreasonableness" (for Australia and the UK) or the substantial evidence requirement and "hard look doctrine" (for the United States); and
vi. Where public comment provisions are prescribed in environmental legislation, procedural fairness may be an obstacle for the Government agency, particularly in the UK and in all likelihood, Australia, to ensure that it meets the legitimate expectation of the public for a full and fair disclosure of information provided to them during the consultation pro cess and upon which their submissions are to be based.

(b) Parties with a Legally Recognised Interest and BATNA Analysis

Standing is the most obvious legally recognised interest that will exist in any environmental conflict. Judicial review legislation (in Australia and the United States) and public law (in the UK) has very broad standing provisions. Similarly, there are also broad standing provisions in these jurisdictions in environmental legislation, including citizen suits in the United States. Consequently, there is generally wide scope for satisfying the standing rules available to parties to have access to the courts and litigation, other than the applicant for a licence or permit under environmental legislation. For these parties their BATNA analysis would need to evaluate the significant differences between the alternative processes for resolving the conflict as there are many possibilities that would be possible as an alternative to negotiation, e.g. litigation, public participation or political lobbying.

A first step in a BATNA analysis would be to assess each alternative process, relative to negotiation, in terms of the extent of shared involvement and joint analysis and control in the administrative decision-making process, i.e. the level of power to parties offered for direct involvement in the administrative decision-making process. In addition, whether the alternative processes incorporated a collaborative approach to fact-finding and problem-solving. Additionally, Fisher *et al.* (1991) advocate that there needs to be sufficient flexibility in the BATNA analysis to ensure the exploration of imaginative solutions. A BATNA analysis would also need to assess whether processes such as litigation, public participation or political lobbying represent an effective alternative to negotiation for resolving an environmental conflict. The following goals of negotiation (Patton, 2005) are particularly relevant in their application as objective criteria to compare each alternative, relative to negotiation:

i. A creative solution that is amongst the best of the available options and which secures as much available value as possible;
ii. A legitimate solution in which the parties have not been taken advantage of;
iii. A commitment that is firm, can be implemented and is sustainable;
iv. An efficient process that reflects good communication; and
v. A process that preserves or enhances the relationships between the parties.

Given the significance of scientific uncertainty in environmental conflicts, a further comparative goal that warrants consideration is the effectiveness of alternative processes for resolving information conflicts. In addition, the extent of any process differences in ensuring that all relevant information is made available and in such a form that it is understood by all parties.

(c) Parties Without a Legally Recognised Interest and BATNA Analysis

Parties that do not meet the standing rules will not have the litigation alternative available to them. In this situation, they may also not have an adequate basis of power to effectively participate in resolving the conflict by negotiation. Consequently, a BATNA analysis may be limited to participatory approaches such as notice and comment under environmental legislation, community consultation and public participation processes — even political lobbying.

However, there is one other possibility to expand the scope of a BATNA analysis. That is, for a party in this situation to consider forming an alliance, or coalition, with another party that meets the standing rules; both parties would need to share a common interest in the environmental conflict. Adopting such a strategy would significantly change the power dynamics by providing a party without standing an opportunity to participate in a wider range of alternative processes for conflict resolution.

SCREENING OF AN ENVIRONMENTAL CONFLICT AND BATNA ANALYSIS

Screening of a conflict is a further dimension of conflict assessment and a BATNA analysis. This aspect of conflict assessment is often referred to as the "gatekeeper role" of the dispute resolver as the assessment of the appropriate process needs to consider not only the subject matter of the conflict, but also the characteristics of the parties themselves.

(a) Subject Matter of the Conflict

Factors that are relevant considerations and which need to be taken into account include:

- Where the nature of the conflict is a test case requiring a legal precedent, litigation is the appropriate process;
- Where there is a lack of decisive case law that leads to uncertainty as to the outcome of litigation, the parties may seek a judicial interpretation of the law. Two alternatives exist where an administrative decision has been made by an agency. First, litigation could be pursued for the court to settle the law and to adjudicate on the merits. Alternatively, the parties could seek a declaration from the court — a remedy in administrative law — to resolve any legal uncertainty as to the meaning of the law. Following the declaration by the court, the parties could then proceed to resolve the factual issues in conflict by negotiation, based on the court's interpretation of the law;
- The existence of intractable factual issues in dispute which effectively limit the scope for compromise and concession-making represents a significant obstacle for negotiation and agreement by consensus; and
- Where the environmental problem is a public interest conflict, negotiation has clear advantages over litigation for conflict resolution — subject to the qualification that there must be an adequate basis of power for the public to participate effectively.

(b) Characteristics of the Parties

Relevant factors that must be considered include:

- The degree of hostility, animosity or polarisation of positions between the parties, especially where it makes negotiating a settlement in good faith problematic. The use of negotiation would be limited in these circumstances;

- The likelihood of there being an inadequate basis of power to ensure a party can effectively participate in negotiations. In these circumstances achieving a fair outcome by negotiation will be limited; and
- Where there is a need, at the very least, to preserve or to enhance the relationship between the parties, negotiation would be the preferred process for conflict resolution.

SHARING POWER AND BATNA ANALYSIS

Where a Government agency believes that a legal challenge could not be made against an administrative decision because parties who oppose the decision lack legal standing, then the outcome of a BATNA analysis may be to ignore the opponents as they would not have access to the courts to litigate the matter. Similarly, where a Government agency believed that there was little risk that a legal challenge against its administrative decision would succeed, it may be inclined not to negotiate with parties having legal standing who are affected by the administrative decision. In these circumstances, litigation as a course of action to defend the administrative decision made may, in all likelihood, be seen as the BATNA.

However, such an evaluation may well change if the agency was less optimistic about the outcome of litigation, e.g. a risk that a judicial review challenge brought against it may succeed. Consider a situation where a BATNA analysis by a Government agency results in a decision to defend a potential civil enforcement action such as a judicial review, rather than to pursue negotiation, as the best alternative. However, the legal challenge by a party affected by the administrative decision ultimately succeeds. The Government agency would find itself in a situation where there is potential for some of the following negative consequences to arise:

- A decline in public trust and confidence in the agency;
- Erosion of the public perception of the agency's capability to perform its statutory duty to protect and manage the environment; and
- Potential embarrassment to the political party in Government.

The alternative, in this example, would have been for the agency to decide, from its BATNA analysis, that negotiation was the appropriate process to resolve the conflict. Consequently, the risk of unsuccessfully defending the legal challenge, and the associated negative consequences for public trust and confidence in the agency, would have been averted. Moreover, positive benefits would also accrue for all parties — and the environment. ADR and interest-based negotiation enables imaginative solutions to be achieved for environmental sustainability — solutions that may not be available through litigation. In addition, in seeking a negotiated agreement, the goal of parties opposing a proposed action that has potential adverse environmental consequences may not necessarily be to block the proposal. Rather, the goal may be to agree to the project proceeding but subject to qualifications. Conditions may be attached to address their needs and concerns about the environment. Inputs provided, in these circumstances, would facilitate the administrative decision-making process. Susskind *et al.* (2000) give a very clear description of the advantages for a Government agency to use negotiation as part of the administrative decision-making process, in this way:

> "[The Government agency] would be able to claim a share of the credit for bringing the conflict to a successful conclusion in a way that saved the cost of litigation, ensured that legal standards of environmental protection were exceeded (voluntarily), and restored some measure of confidence in the Government."

Where a Government agency considers the potential threat of some form of civil enforcement action as real, rather than speculative, and that there is a reasonable likelihood that there will be a risk in successfully defending such an action, the threat of litigation becomes the catalyst for the agency to consider sharing its power in the administrative decision-making process. The pre-condition for managing and resolving an environmental conflict through sharing power and negotiation is knowledge of legal rights contained in administrative and environmental law.

FINDING SOLUTIONS FOR ENVIRONMENTAL CONFLICTS THROUGH NEGOTIATION: UNIFYING PRINCIPLES

Susskind and Hoben (2004) have proposed seven principles to guide collaborative efforts for decision-making to manage development in metropolitan areas where the issues involve multi-objective considerations — such as economics, open space preservation and infrastructure — which require balancing in the decision-making process. Significantly, their principles are linked to established consensus-building theory. Environmental conflicts are problems in sustainability; resolution of sustainability conflicts requires a consideration and balancing of multiple objectives: ecological, economic, social and cultural. Consequently, the guiding principles of Susskind and Hoben also represent unifying principles for the resolution of environmental conflicts where consensual agreement is sought using a mutual gains approach. Therefore, the principles, as defined and described by Susskind and Hoben, may be directly applied to the resolution of site-specific and policy-level environmental conflicts and, in this context, have been summarised as follows:

i. *From the outset, all efforts musts be directed at identifying all relevant parties and then inviting them to participate*: This principle for public participation has long been recognised to be of critical importance — but problematic. Exclusion of parties having a relevant interest in the conflict may not only lead to a non-viable outcome, but also have the potential to undermine the legitimacy of any consensus-building. Where a collaborative approach is used to manage and to resolve conflict, great care must be taken to ensure all relevant parties holding competing interests are included.
ii. *The conflict management and resolution process should be transparent and accountable to all parties, Government agency administrative decision-makers and the public at large*: To increase the effectiveness of collaborative decision-making, the goals of conflict management and resolution should be made broadly known in terms of who is involved, an awareness of the level of power available to the parties, and the process in which they will all be involved. Any perception that the process is not transparent and accountable may lead to a lack of public trust and confidence in the process, and possibly in the Government agency. The commitment of the parties directly involved in the process also may diminish if there is any perception of manipulation by the agency.
iii. *The process of shared responsibility and joint action in finding solutions for environmental conflicts should give all parties access to the available information as well as an understanding of the scientific/technical data*: Creative options for mutual gain can best be developed when all the parties have available to them scientific information that is both relevant and reliable. Joint fact-finding involves the independent scientific experts of the parties using objective criteria in an open and transparent process to assess the scientific/technical data. This assessment is the foundation for the parties to find solutions to the conflict.
iv. *Consensus should seek to maximise joint gains, in creative ways, from a range of options that recognise existing legal and institutional constraints*: The goal of collaborative efforts ought to be to create as much value as possible whilst partitioning that value in a manner that is perceived by all parties to be fair, efficient and feasible. Barriers may be imposed by legal, institutional and organisational structures. However, significant gains can be found by developing a number of sustainable

land use options for mutual gain, for a range of issues, and then encouraging parties holding different values to trade across the options.

v. *The question of how a negotiated agreement arrived at will be implemented needs to be addressed by all parties as part of the conflict resolution process*: Particular attention must be given to clarifying the procedures required to ensure that the agreement will be implemented, by whatever formal means are appropriate. This step will reflect the level of power that the parties have in the administrative decision that is to be made. Ensuring that the Government agency administrative decision-maker acts on the outcome of the conflict resolution process will be facilitated where implementation efforts involves all parties to the conflict resolution process.

vi. *The Government agency convenes the conflict management and resolution process and reserves the right to ensure any and all final administrative decisions are consistent with legislative obligations*: An agency engaged in collaborative decision-making is not abandoning its power to make the final administrative decision. Rather, it is willing to share power to find a solution in a way that is transparent and responsive to the needs of parties holding a recognised or relevant interest in the conflict. Ultimately, the agency is accountable for the final administrative decision being consistent with legislative obligations.

vii. *Professional neutrals should be used to organise and to facilitate the conflict assessment, management and resolution process*: External and independent professional dispute resolvers, with no bias towards a particular outcome, have a key role in assisting the parties to find their own solutions to the conflict, e.g. by facilitating parties to develop sustainable land use options for mutual gain; ensuring all relevant and reliable scientific information is available and is understood; providing an atmosphere for negotiations that is conducive for achieving fair and positive outcomes; and facilitating land use options being created that take into account legislative obligations, where the law is not in dispute.

ADR PROCESSES AND ENVIRONMENTAL CONFLICTS

Based on the analysis of Sourdin (2005), ADR processes can be classified as either facilitative, advisory or determinative. A "facilitative process" involves a dispute resolver assisting the parties to manage the process of conflict resolution, but without having an advisory or determinative role. On the other hand, an "advisory process" provides the dispute resolver with the role of investigating the conflict in order to provide advice on the facts and possible outcomes. "Determinative processes" can be distinguished from advisory processes as the dispute resolver investigates the conflict and makes a determination which may be potentially enforceable. This distinction is important in choosing an ADR process to resolve an environmental conflict as different ADR processes need to be used for the different stages of conflict assessment, management and resolution. Determinative ADR processes, such as arbitration, where the determination made by the dispute resolver is binding on the parties, would have limited application in managing or resolving environmental conflicts where a collaborative approach to fact-finding and problem-solving was adopted.

Furthermore, the scope of different ADR processes needs to be reflected in the skills required for a dispute resolver. ADR process skills are one essential foundation. Depending on the ADR process, subject matter expertise in science and/or law is equally as important as ADR process skills for the stages of conflict assessment, management and resolution.

(a) BATNA Analysis and Power-Sharing: Early Neutral Evaluation

Early neutral evaluation is an advisory process. It is undertaken early on in litigation before much discovery is done. The purpose is to provide the parties with an early and frank non-binding evaluation of the legal merits of their case, e.g. the likely outcome of a judicial review challenge. Early neutral evaluation has been defined by the Law Society of New South Wales, Australia (cited by Duncombe and Heap, 2004) as: "*a process whereby parties obtain a non-binding, reasoned evaluation of their case on its merits by a neutral third party acting as an evaluator (Australasian Dispute Resolution, 2004)*".

(b) Conflict Assessment: Facilitative Mediation

Facilitation is the primary process used for conflict assessment. Mediation may be undertaken in a passive manner where the dispute resolver has no advisory or determinative role. This process is referred to as "facilitative mediation" (Sourdin, 2005). Boulle (2005) defines facilitative mediation in the following way:

> "[A model for mediation that has as its] main objective to avoid positions and negotiate in terms of parties' underlying needs and interests instead of their strict legal entitlements. [The mediators' main role is to] conduct the process, maintain a constructive dialogue between the parties, enhance negotiation efforts and encourage settlement".

(c) Conflict Management: Independent Expert Appraisal

Independent expert appraisal is an advisory ADR process. The expert appraiser is essentially a fact-finder and evaluates the disputed facts and provides a non-binding opinion or non-judgemental advice to the parties on how each issue in dispute may be resolved. The basis of the expert appraiser's non-binding opinion is information provided by the parties complemented by further research undertaken by the expert appraiser. The non-binding opinion may suggest options for the parties to consider for negotiation and compromise.

(d) Conflict Management: Evaluative Mediation

Mediation may also be undertaken with an advisory role where the dispute resolver has an active role in evaluating the conflict and providing advice to the parties in order for the parties to find their own solution by consensus. This process is referred to as "evaluative mediation". Evaluative mediation is a process where the dispute resolver evaluates the merits of the conflict and provides suggestions or recommendations on how the conflict could be resolved (Sourdin, 2005).

(e) Dispute Resolver Skills

The accepted body of ADR knowledge recognises the need for an independent third party, acting as the dispute resolver, to be of paramount importance to assist the parties to find their own solution to a conflict by mutual agreement. The involvement of a neutral, professional dispute resolver has advantages for ensuring that the interests of all affected parties are effectively taken into account in the processes of conflict management and resolution. Another advantage is the ability to raise creative options for mutual gain that can be considered by the parties to resolve their conflict.

The underlying features of any environmental conflict emphasise the significance of cross-disciplinary knowledge in science and law that must be considered in terms of finding solutions for conflicts over sustainability. Achieving greater cross-disciplinary knowledge must also be considered in the context of knowledge of ADR processes as different ADR processes need to be undertaken at different stages of the environmental conflict. What does this mean for cross-disciplinary collaboration and the resolution of environmental conflicts? A dispute resolver specialising in the resolution of environmental conflicts — especially environmental regulatory or procedural conflicts — must have ADR process skills as well as expertise in the subject matter of the conflict, i.e. administrative and environmental law and/or environmental science. For example, the dispute resolver undertaking an early neutral evaluation would require expertise in law and ADR process skills. A dispute resolver undertaking an independent expert appraisal would require expertise in science and ADR process skills.

A MODEL FOR MANAGING AND RESOLVING ENVIRONMENTAL CONFLICTS BY NEGOTIATION

Adoption of the accepted body of knowledge for effective public participation (Institute for Participatory Planning, 1981), principled negotiation (Fisher *et al.*, 1991) and the mutual gains approach to negotiation (Susskind *et al.*, 2000) is essential for resolving environmental conflicts and so is the foundation for a model for managing and resolving environmental conflicts (Figure 10.1). The model focuses initially on conflict assessment and then extends to conflict management and resolution. Flexibility, depending on the stage of the conflict, is a paramount consideration in choosing the appropriate ADR process. Different ADR processes are used that are specific to the needs of each stage. The model focuses on cross-disciplinary collaboration by linking the appropriate skills required for a dispute resolver to the sequential stages of conflict assessment, conflict management and conflict resolution. Undertaking all the sequential stages of conflict assessment, management and resolution may be necessary in order to negotiate an agreement to resolve an environmental conflict. However, on a case-by-case basis, there is also scope to consider the potential application for the use of a specific stage as a discrete part in itself, e.g. the conflict management phase, where information conflicts need to be resolved in order to facilitate the integrity of the administrative decision-making process.

(a) Problem Identification and Characterization

The ignition phase emerges when the public first become aware that a proposed action may have the potential to cause environmental harm. Characterisation of the potential problem should focus on the type of administrative decision-making process used by the Government agency. The decision may be made under environmental protection legislation, where the aim of regulatory control to prevent harm to the environment or human health is through an administrative decision to issue a permit or licence before the action can proceed. Alternatively, the legislative approach to prevent harm to the environment or human health may be procedural, e.g. the EIA process. Legislative procedures are set out that must be complied with before an administrative decision can be made as to whether the proposed action should be approved.

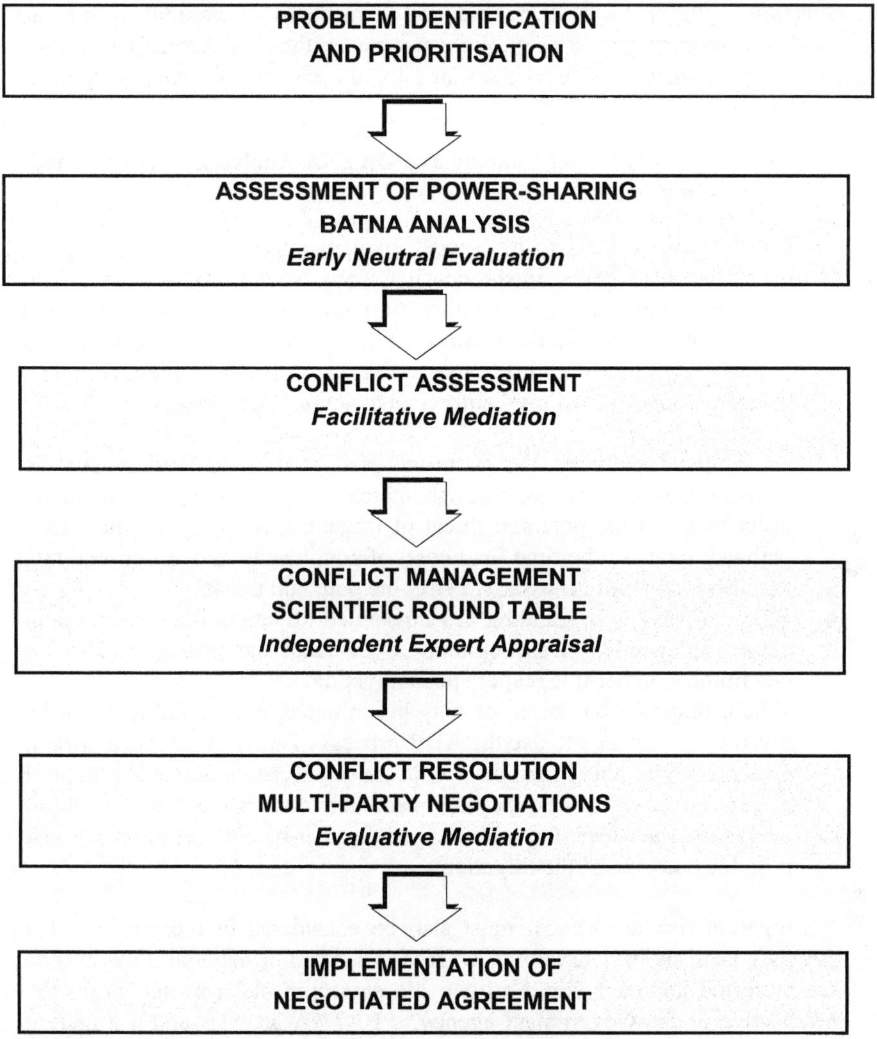

Figure 10.1 Procedures and processes for managing and resolving environmental conflicts by ADR and structured negotiation

Depending on the specific legislation under which the administrative decision is to be made there may be a mandatory requirement for the Government agency to seek public comment on the proposed action. Public comment provisions have implications for the nature and extent of public participation.

(b) Assessment of Power Sharing and BATNA Analysis: Early Neutral Evaluation

Confronted with a scenario of a legal challenge from parties that have a legally recognised interest in the conflict, such as standing, a Government agency could move along a number of pathways to evaluate the risks associated with litigation. Some form of litigation risk assessment would indicate whether there would be a reasonable likelihood for a judicial review challenge, or some form of civil enforcement action, to succeed:

- A legalistic pathway, by auditing each step of the administrative decision-making process, through ongoing inputs of legal opinion, in order to avoid the potential threat of litigation. Not only would such a pathway increase the time (and cost) of administrative decision-making, but also it may not necessarily offset the potential threat;
- Where a party with standing has threatened litigation, the agency could obtain independent legal opinion to evaluate the potential claim to determine whether it is real or speculative; or
- Where litigation has been, or may be, initiated, consideration could be given by the agency to use the ADR process of *early neutral evaluation.* The parties to the litigation would be involved in the early neutral evaluation process. The dispute resolver convening the early neutral evaluation provides the parties with a non-binding opinion on the possible outcome of the litigation.

Litigation risk assessment must also be considered in the context of a BATNA analysis in determining whether litigation or negotiation would be the preferred approach for resolving the environmental conflict. From the perspective of the Government agency, a BATNA analysis and a litigation risk assessment provide the basis for determining whether to exercise its discretion and be willing to share power to use negotiation as the basis of the administrative decision-making process. Notwithstanding that a Government agency may decide that it is willing to share its power to use negotiation to resolve the conflict, this would be a preliminary decision. The agency would need to consider whether the alternative of negotiation is feasible. Conflict assessment, the first step in the process of conflict resolution, needs to be undertaken.

(c) Conflict Assessment: Facilitative Mediation

Conflict assessment is a well-established procedure used to evaluate, amongst other things, whether there is a reasonable likelihood of managing or resolving an environmental conflict by negotiation. The conflict assessment needs to be undertaken by a professional dispute resolver, independent of all parties involved in the conflict. Given that an environmental conflict will involve complex and numerous scientific facts, the dispute resolver needs to have both ADR process skills and expertise in the scientific subject matter of the conflict. Most of the phases of conflict assessment would use the ADR process of facilitative mediation.

Susskind and Thomas-Larmer (2000) and Schmeuli (2003) consider that the benefits of conflict assessment include (i) the identification of all relevant parties as well as their willingness to negotiate in good faith, (ii) consolidation of parties into coalitions (or groups) having similar interests or issues and (iii) the identification of issues in dispute and common ground.

Identification of parties

The agency must ensure that all efforts to identify all parties that have an interest in the administrative decision have been made. There are two possibilities, which can be distinguished in terms of the extent of involvement of parties in negotiating agreement. First, the number of parties involved in resolving the conflict could be restricted where the Government agency required the parties to establish they had legally recognised interest, such as standing, to pursue a civil action in relation to the administrative decision.

However, there is another possibility for public interest environmental conflicts — one of wide participation. Parties who do not satisfy the standing rules, but whose interests may be affected by the administrative decision, may also warrant consideration for being involved in resolving the conflict. Where the environmental legislation under which the administrative decision is made prescribes requirements for public notice and comment, and particularly where the environmental problem is a public interest conflict, the situation changes. Specifically, for parties who have made a submission in response to a public notice seeking comment on the proposed action. Legal principles in the UK (and an arguable case in Australia), that they are entitled to have a legitimate expectation, as a matter of law, for the information disclosed for public comment to be full and fair, would be a relevant consideration for the agency for the process of identifying parties. For a public interest environmental conflict, in all likelihood, their numbers would be large.

Party numbers

Whilst there is considerable debate as to the optimal number of parties that can ensure effective involvement in negotiating solutions, Comly French (1993) suggests 15 as ideal, with an upper limit of 20–25. There is scope for reducing the number of parties in public interest environmental conflicts by forming alliances or coalitions of interests.

The coalition of interests concept and multi-party environmental conflicts

A coalition of interests is composed of parties having common interests, goals or aspirations or parties likely to be affected by the administrative decision in a similar manner (Box 10.1). There are three broad groups of coalitions of interests that are generally recognized, based on the impacts of the administrative decision on the following interests: “economic”, “special” and “community” (Comly French, 1993).

Box 10.1 Forming coalitions (or alliances) of common interests in multi-party environmental conflicts (based on Comly French, 1993)

Coalition of Economic Interests

Parties with an economic interest in the administrative decision to be made.

Examples of specific interest groups within the coalition could be the party seeking the licence or project approval, local authorities, landholders.

Coalition of Special Interests

Parties with a special interest in the administrative decision to be made, but which have no immediate financial stake.

Examples of specific interest groups within the coalition could be Indigenous communities and cultural heritage, conservation groups, public health organisations.

Coalition of Community Interests

Parties living in the area that have to live with the outcome of the administrative decision.

Examples of specific interest groups within the coalition could be resident action groups that focus on discrete local issues such as public health, safety, amenity or social impacts.

The essential features of the concept can be summarised as follows:

- Parties that have been identified as having an interest that may be affected by the administrative decision are asked, during the conflict assessment stage, to assign themselves to one coalition of interest;
- Where there is overlap in choosing a specific coalition, only one coalition which represents the parties' substantial interest, i.e. a coalition of economic, special or community interest, must be chosen;
- Next, there is a subdivision within each of the three coalitions of interest. Parties are placed within a specific interest group within each coalition that best reflects the specific issues which have been raised in their public submissions, e.g. Government, primary production, Indigenous community, town planning, natural resources, cultural heritage, youth or local community. This step is initially undertaken by the dispute resolver based on the analysis of interests as contained in submissions made during the public comment period;
- The number of parties that are represented in each specific interest group will probably be unequal. However, the paramount consideration is to ensure all relevant interests are represented in the specific interest groups assigned — not the number of parties that comprise each specific interest group;
- The parties in each specific interest group, within each coalition of interest, elect their representative having their full authority to participate, on their behalf, in negotiations throughout the conflict management and resolution process, including entering into agreements;
- The role of the representative extends to include responsibilities such as scheduling regular briefing sessions before commencing any negotiations on issues relevant to the specific interest group, to keep track of all inputs from their specific group to ensure they were properly considered in negotiations and to make all parties in their specific interest group aware of all matters in negotiation that are relevant to them. Canadian experience indicates that adherence to these responsibilities is crucial for ensuring the success of this technique (Stuart, 1992); and
- Moreover, forming coalitions or alliances would facilitate the scope for sharing costs, as larger numbers of affected parties sharing a common interest would be involved in public interest conflict resolution.

Balanced representation of interests

The model is based on a balanced representation of interests of all parties affected by an administrative decision and having a specific interest in the conflict. Balanced representation ensures that the process for negotiating agreement is procedurally fair. All representative interests have an equal opportunity to ensure that their view is properly taken into account in any solution reached by mutual agreement. Balanced representation in the model is achieved through the concept of coalition of interests. The goal of balanced representation can be achieved when the number of specific interest groups, within each coalition of interest, is the same. For example, assuming 15 as the ideal number of representative interests to be involved in negotiating solutions, 5 specific interest groups within each of the 3 coalitions of interests would result in a balanced representation.

Outcome of conflict assessment

The dispute resolver would need to provide conclusions to the Government agency on participation issues such as: whether all interested parties had been identified; whether there was a reasonable likelihood that a negotiation round table could be convened with balanced representation; and whether the negotiation round table would adequately represent the interests identified as being affected by the administrative decision. The outcome should also include conclusions from a BATNA analysis as well a screening of the conflict, e.g. by considering the subject matter of the conflict and the characteristics of the parties themselves. A recommendation can then be made to the Government agency as to whether the preliminary decision of the agency to share power with the parties to resolve the conflict by negotiation is justified.

Where the recommendation is positive, the dispute resolver should provide a proposed list of parties together with a draft structure assigning them into proposed coalitions of interests and specific interest sub-groups within each coalition. A preliminary scoping exercise should also be undertaken in which the dispute resolver provides a draft that identifies the key issues in dispute as well as common ground. This scoping exercise would be based on a review and analysis of the public submissions received.

(d) Conflict Management — Preliminary Matters: Facilitative Mediation

The initial phase of conflict management should be convened by the independent dispute resolver who had undertaken the conflict assessment. The parties would be the Government agency and the interested parties identified by the dispute resolver in the conflict assessment. First, the parties need to endorse the proposed list of parties together with the proposed structure assigning them into coalitions of interests and specific interest sub-groups within each coalition of interest — the outcome of the conflict assessment. The parties have the liberty to vary how they have been assigned to a coalition or to a specific interest group. They may also identify alternative interest groups which may be more appropriate than the specific interest groups proposed from the conflict assessment outcome. When the coalition of interests and the specific interest groups have been accepted and finalised, the parties in each specific interest group within each coalition elect their representative who has their authority to participate in the round table negotiations, on their behalf. Consequently, the parties in the round table negotiations would be the representatives of each specific interest group for each coalition plus a representative of the Government agency ("the round table representatives").

The preliminary scoping exercise undertaken by the dispute resolver in the conflict assessment stage should be the first subject of round table negotiations. The key issues in dispute plus any common ground that were identified should be reviewed by the round table representatives, revised as appropriate, prioritised and then endorsed. The agreed list of issues becomes the focus for the next phase of conflict management.

(e) Conflict Management — the Scientific Round Table: Independent Expert Appraisal

The scientific round table is based on the ADR process of *independent expert appraisal.* The only issues addressed at the scientific round table are factual issues in which divergent opinion exists. The dispute resolver must have ADR process skills as well as expertise in the scientific subject matter of the conflict. The parties at the scientific round table are scientific experts, nominated by each specific interest group, to act for and to represent them.

Data (information) mediation

The environmental issues that have been scoped and agreed to by the round table representatives become the foundation for the data mediation. A list of scientific materials, e.g. published articles, reports, experimental studies, is compiled at the scientific round table to address the scoped issues. Each scientific expert acting for a specific interest group may identify any relevant materials. In addition, the dispute resolver may also identify materials to complement the list compiled by the scientific experts. The final list of materials becomes the common database used in the next step — collaborative joint fact-finding at the scientific round table.

Scientific round table

The purpose of the scientific round table is for the scientific experts to reach agreement by consensus for each disputed factual issue identified in the scoping exercise. To avoid information conflicts arising from different interpretations of the common database, the scientific round table must adopt an underlying element of principled negotiation (Fisher *et al.*, 1991); i.e. to discuss and agree to objective criteria that can be used to evaluate each disputed issue, using the common scientific database. Where consensus cannot be reached on any issue, the dispute resolver provides a non-binding opinion based on an objective and impartial analysis of all relevant and reliable scientific data. The purpose of the non-binding opinion is to facilitate the experts to reach consensus on the disputed issue.

Outcomes of the scientific round table

Negotiation of solutions to resolve the conflict do not commence until all relevant scientific and technical information related to the key environmental issues that have been scoped in the preceding phase of conflict management have been fully disclosed and then evaluated. The dispute resolver is required to prepare a summary of outcomes from the scientific round table for the round table representatives who will be involved in finding solutions, by negotiation, to resolve the conflict. Consequently, it is important that the conclusions reached by the scientific experts are communicated to the round table representatives by the dispute resolver in a form that they understand. The summary of outcomes would include:

- Conclusions on disputed issues where agreement is consistent with all relevant and reliable scientific data and/or scientific opinion;
- Where agreement cannot be reached by the experts on a disputed issue the non-binding opinion of the dispute resolver would be provided;
- Areas of scientific uncertainty for a specific issue, including where there is a lack of information, must be identified — especially where it would lead to conclusions being seen as speculation; and
- A number of alternative pathways may be suggested where the available scientific information associated with a specific issue in dispute is either uncertain or missing. For example, whether consideration should be given to applying the precautionary principle and whether there is a need for the precautionary principle to be reflected in any conditions that might be imposed. Another alternative would be to consider a contingent agreement that provided for an opportunity to review and vary the commitments in the agreement when the available scientific information changed. Finally, a recommendation for the need for further scientific studies could be made. The scientific round table would prepare a summary of agreed finding of facts from the common scientific database as well as any underlying assumptions upon which the study was based. In addition, the scientific round table would draft the terms of reference for the proposed study and agree on appropriate experts or organisations who should undertake the study.

Knowledge power and the scientific round table

The collaborative joint fact-finding approach of the scientific round table is based on the use of a common database to address the factual issues in dispute identified by the parties and their evaluation using objective criteria prepared by the scientific experts. Such an approach ensures that there is an adequate basis of power for all parties to effectively participate in the conflict. The outcomes of the scientific round table process become the foundation for the parties to share scientific knowledge equally in the final stage, conflict resolution, in circumstances where there has been a full and fair disclosure and evaluation of all relevant information for all parties.

(f) Conflict Resolution — Multi-Party Round Table Negotiations: Evaluative Mediation

The final stage is a collaborative process of joint problem-solving involving the representative of each specific interest group within each coalition of interest, a representative of the Government agency that must make the administrative decision and convened by an independent dispute resolver. Factual issues are not in dispute as this stage has its foundation in the scientific round table outcomes. Where the negotiated agreement needs to take into account legislative obligations, the dispute resolver must have subject matter expertise in the law as well as ADR process skills.

The joint problem-solving approach to conflict resolution is structured on two fundamental elements of principled negotiation: interest-based negotiation and creating a number of options for mutual gain. The interest-based approach to negotiation in multi-party environmental conflicts requires agreement to be reached by consensus. There is also a need for the parties to undertake negotiations in good faith.

Interest-based negotiation

Exploring options for mutual gain requires parties at the round table identifying their essential interests as well as understanding the interests of the other parties. The options need to be considered against some agreed to standards such as fairness, reasonableness and consistency with matters that must be considered under the legislation for approving the issue of a licence or a permit. Canadian experience (Stuart, 1992) indicates that the following benefits arise from interest-based negotiation:

- Assisting the parties to understand and to accept the legitimacy of everyone's interests. In turn, to promote a new commitment and creative energy for achieving an agreement.

- Enabling the parties to develop a firm sense of the interconnectedness between all parts of an agreement being developed for mutual gain. The alternative is to negotiate discrete parts of the agreement in isolation. This may be problematic for achieving integration between all parts of any agreed outcome.
- Interest-based negotiation avoids differences in the positions of the parties. Negotiation on the basis of positions leads to situations which defy any prospects for agreement.

Consensus decision-making

The goal of consensus decision-making is to encourage joint problem-solving. In multi-party negotiations, considerable thought has to be given to how agreement is to be reached between all parties. For example, if agreement is to be achieved by a unanimous decision, there is the potential for an agreement not to emerge as it would be possible for any one party to use their right to veto the decision and hold out if they so desired?

The alternative to a unanimous agreement is consensual agreement. The generally accepted meaning for consensus is reaching a decision that all parties can live and abide with (Pritzker, 1995). A very good example of applying the concept of consensual decision-making to multi-party environmental conflicts is contained in the following groundrules developed in the "*Northeast British Columbia '2005' Initiative*" as cited by Cormick *et al.* (1996):

- Consensus does not mean total agreement on every part of a decision, but all parties must be willing to accept the overall option for mutual gain;
- Where a party withholds agreement on an issue, that party is given the opportunity to explain how their interests are adversely affected or how the proposed agreement fails to meet their interests. Alternatives must be proposed by the party withholding agreement which must then be considered by the other parties in terms of how all interests may be met;
- Prior to reaching final agreement in a multi-party environmental conflict, the round table representative of each specific interest group, for each coalition of interest, would need to take the agreement back to the members of their interest group for ratification. Similarly, the representative of the Government agency at round table negotiations may also need to seek ratification from a higher decision-making authority;
- Where consensus is reached it is assumed to be binding for the entire negotiated agreement; and

- Where consensus is not reached, the rights and obligations of each party will not be prejudiced as a consequence of participating in the process.

Negotiating in good faith

Finding solutions for environmental conflicts requires all parties to negotiate in good faith. This raises the question of what does good faith mean? In turn, how can the presence of good faith be assessed? This issue was addressed by the National Native Title Tribunal in Australia [in *Western Australia v Taylor* (1996) 134 FLR 211] in the context of negotiations for land use conflicts involving Indigenous communities. The Tribunal identified 18 indicators that could be used to evaluate the question of good faith. Some examples for indicators that applied include:

- *The conflict assessment stage*: Unexplained failure to communicate with the other parties within a reasonable time;
- *The conflict management stage*: Failing to respond to reasonable requests for relevant information within a reasonable time;
- *The conflict resolution stage*: Adopting a rigid non-negotiable position; a failure to make counter-proposals; a refusal to incorporate statutory provisions into an agreement; shifting position just as agreement seems in sight.

The negotiated outcome

The outcome of the round table negotiations is to reach consensual agreement on whether a permit or licence or approval should be given for a proposed action that has potential adverse environmental consequences. By trading across issues which the parties value differently, the goal would be to agree to an option which would leave everyone better off than they had the right to expect, compared with the most likely outcome arising from litigation (Susskind *et al.*, 2000). The option agreed to could possibly be a new or "preferred option" that had been constructed based on the best features of the original options. The option agreed to would also incorporate a number of interconnected parts, e.g. conditions to ensure sustainability being achieved as well as measures to prevent potential environmental harm. In this regard, the negotiated agreement would need to ensure that any conditions to ensure compliance with sustainability are clearly expressed and capable of being effectively implemented through an environmental audit, e.g. ecological monitoring, an environmental management plan, performance bond and measures for managing risk where scientific uncertainty exists.

Post-negotiation and implementation

The catalyst for this model for resolving environmental conflicts arises from a BATNA analysis. The Government agency decides that it is willing to share its power in the administrative decision-making process with parties having a recognised interest in the conflict; negotiation is seen as the most appropriate process for conflict resolution. The agency would also recognise that there were a number of available options for managing and/or resolving conflict in the administrative decision-making process under environmental legislation aimed at preventing harm to the environmental or human health. In circumstances such as these, where a negotiated agreement is the outcome, Pritzker (1995) states that the agency would be ultimately responsible for ensuring that the administrative decision was consistent with the legislative requirements for approval of the proposed action. However, concern may arise because of the existence of a standard where a Government agency reserves the right to make any, or all, final decisions to ensure that the negotiated outcome is consistent with the law. Could the implementation of a negotiated agreement be problematic if the administrative decision did not follow the outcome reached in a collaborative decision-making process? That is, could adherence to such a standard produce a situation where a negotiated agreement arising from a collaborative approach to decision-making does not end in a binding contract? Rather, that the negotiated outcome may only be advisory?

However, Susskind and Hoben (2004) provide a positive perspective for this dilemma. They contend that where there is clarity from the outset about the relationship between the consensus-building and the ultimate administrative decision-making processes, "*there should be no surprises*". However, they do acknowledge that there may be some instances beyond the control of the parties — and the administrative decision-makers themselves, that may undermine implementation. In such a situation, a failure to implement a collaborative decision needs to be counter-balanced against the prospects of a judicial review challenge being real or speculative.

In this regard, public interest environmental conflicts can be distinguished from private matters. In private matters, the parties mediating are generally not part of the administrative branch of Government; judicial review would not be available if the accountability of administrative officials and Government agencies was not in issue (Boulle, 2005). However, this situation may not exist where a Government agency shares its power with interested parties in order to reach a negotiated agreement for managing or resolving an environmental conflict — ultimately ending in an administrative decision made by the agency. This decision will affect the rights of the interested parties — at least those with a legally recognised interest in the conflict who satisfy the standing rules. In addition, the parties with affected rights may extend to those who have responded to the public comment provisions mandated by statute. In issue, in the latter situation, would be the question of procedural fairness (for Australia and the UK) and, possibly, procedural due process (for the United States). In particular, what had the Government agency committed itself to, by promise or practice, that the parties involved in the collaborative decision-making process were entitled to hold from the outset, as to the relationship between the processes of consensus-building and the final administrative decision?

There is also another possibility for a judicial review. That is, consideration for a challenge of the administrative decision on the grounds of the "no-evidence rule/unreasonableness" (for Australia and the UK) or the substantial evidence rule and the "hard look doctrine" (for the United States). Therefore, an administrative decision made by a Government agency that sets aside the advisory outcome arising from a collaborative decision-making process, raises the potential threat of a new judicial review challenge. Clearly, such a result will not benefit anyone. The scope for the future involvement of the public in collaborative decision-making would be adversely affected. Public trust and confidence in the Government agency may deteriorate.

The potential use of judicial review, in the same environmental conflict, as both a trigger to share power to facilitate the use of a collaborative approach for managing and resolving an environmental conflict — as well as a check and balance on the integrity of the final administrative decision made following a negotiated agreement — highlights its *utility* to address any abuse or misuse of power by a Government agency.

CONCLUSION

The procedures for assessing, managing and resolving environmental conflicts described in the model reframe the role of science and law for finding solutions for environmental problems; problem-solving has a cross-disciplinary dimension. The roles for legal and non-legal environmental professionals as dispute resolvers are different, but complementary. Their roles reflect the scope and object of the ADR processes used at different stages of the development of an environmental conflict. However, both law and science have shared goals: to find solutions for environmental conflicts through creative problem-solving approaches based on ADR, interest-based negotiation and consensus-building.

The model is based on the sequential stages of conflict assessment, management and resolution and has general application for environmental conflicts. For environmental conflicts that are to be resolved by negotiation, external to and independent of the courts, the parties themselves have the flexibility to decide on the procedures that will be used to resolve their specific conflict. Procedures for ADR and negotiation are not prescribed in these circumstances. Therefore, it may be prudent to compare the procedures and processes in this model for resolving an environmental conflict against other pathways for ADR and negotiation, against objective criteria for achieving the goals of negotiation, e.g. Patton (2005).

For parties using court-annexed ADR to resolve a specific environmental conflict, the procedures used in the model will not, in all likelihood, be mandated. Consequently, there may be some justification for the parties involved to consider whether adoption of some, or all, of the conflict management and resolution procedures may be warranted — particularly if their goal was to offset the concerns expressed by Susskind (2005) over the use of court-annexed ADR and consensus-building.

Bordone *et al.* (2005) provide a stimulating analysis on the challenges and opportunities confronting dispute resolution in the 21st century. Notwithstanding the enormous progress made in the theory and practice of ADR, the authors argue that conflict management and resolution is at the crossroads. One of the challenges they raise reflects the approach taken in the model: resolving multi-party environmental conflicts based on cross-disciplinary collaboration and the use of knowledge and resources. In a small way the cross-disciplinary approach taken in developing this model for managing and resolving environmental conflicts may "*encourage the kinds of cross-disciplinary collaboration that are so critical to the continued development of our field*" (Bordone *et al.*, 2005).

Common terminology adopted

Alternative Dispute Resolution ("ADR") and Negotiation
Negotiation is the primary form of dispute resolution and the foundation for a wide range of ADR processes, such as mediation, independent expert appraisal and neutral evaluation. Any discrete reference to the processes of "*ADR*" or "*negotiation*" should be interpreted as an inclusive one, as it combines both elements. Where a specific form of negotiation is relevant, e.g. interest-based negotiation, the reference to ADR is qualified to indicate the process as ADR and interest-based negotiation.

Court-Annexed ADR
ADR processes incorporated as part of the court system and legislative procedures that are used prior to, or during, court proceedings are commonly referred to as "court-annexed" ADR. Other variants of this term include "court-based", "court-connected" and "court-related" ADR. Where it does not end in agreement, the matter then proceeds to a court hearing.

ADR processes may also be undertaken to resolve conflict independently of, and with no connection or relation to, litigation initiated in a particular court system. ADR processes used in these circumstances are termed "non-court-annexed ADR".

Ex parte
A legal term that is used in the title of a reported case to point out that the proceedings involve one party only, i.e. the application before the court is that of the party whose name follows this term.

Government Agency
Administrative decisions for environmental protection and management are made by Government, but the terminology used to define the bodies involved in Australia, the UK and the USA is diverse, e.g. departments, local authorities, councils, planning authority, county, shire, etc. Consequently, any general reference to the term "Government agency" is a reference to the multitude of Government bodies having the statutory power to make administrative decisions in relation to the environment.

Litigation

There is a clear distinction in the legal process between the adversarial system of Australia, the UK and the USA and the inquisitorial system of most of continental Europe, Japan and South America. Therefore, any general reference to the term "*litigation*" relates to the use of the courts to adjudicate on environmental conflicts, independently of the system used. Where specific features are applicable to the adversarial system, the reference is made to "*adversarial litigation*".

Party

The wide range of processes used to address environmental conflicts results in a broad spectrum of terms used in the literature to refer to those involved: "applicant", "appellant", "respondent", "plaintiff", "defendant", "litigant", "participant", "stakeholder" and "consultee". The standard term used throughout the book is the generic term "*party*" as a neutral term that is applicable, independent of the process used.

Threatened Species

The term "*threatened species*" is used throughout the book as a general term to incorporate all the various conservation categories as defined and applied in environmental legislation, such as "endangered', "threatened", "vulnerable", "endemic", "rare" and "conservation-dependent".

Glossary of acronyms

AAT Act: *Administrative Appeals Tribunal Act (Australia)*
AD(JR) Act: *Administrative Decisions (Judicial Review) Act (Australia)*
ADR: *Alternative Dispute Resolution*
APA: *Administrative Procedure Act (USA)*
APHIS: *Animal Health and Plant Inspection Service (USDA)*
BATNA: *Best Alternative to a Negotiated Agreement*
CERCLA: *Comprehensive Environmental Response, Compensation and Liability Act (USA)*
CFR: *Code of Federal Regulations (USA)*
CWA: *Clean Water Act (USA)*
EA: *Environmental Assessment*
EEC: *European Economic Community*
EIA: *Environmental Impact Assessment*
EIS: *Environmental Impact Statement*
EPA: *Environmental Protection Agency*
EPBC Act: *Environment Protection and Biodiversity Conservation Act (Australia)*
ES: *Environmental Statement*
ESA: *Endangered Species Act (USA)*
ESD: *Ecologically Sustainable Development*
EU: *European Union*
FIFRA: *Federal Insecticide, Fungicide and Rodenticide Act (USA)*
FOI: *Freedom of Information*
FOIA: *Freedom of Information Act (USA)*
FONSI: *Finding of No Significant Impact (NEPA, USA)*
GMOs: *Genetically Modified Organisms*
GT Act: *Gene Technology Act (Australia)*
IUCN: *International Union for Conservation of Nature and Natural Resources*
NEPA: *National Environmental Policy Act (USA)*
PPC Regulations: *Pollution Prevention and Control (England and Wales) Regulations (UK)*
RCRA: *Resource Conservation and Recovery Act (USA)*

SDWA: *Safe Drinking Water Act (USA)*
SEA: *Strategic Environmental Assessment*
SLAPP: *Strategic Litigation Against Public Participation*
TCP Act: *Town and Country Planning Act (UK)*
TCP Regulations: *Town and Country Planning (Environmental Impact Assessment) (England and Wales) Regulations (UK)*
ToSCA: *Toxic Substances Control Act of 1976 (USA)*
UNCED: *United Nations Conference on Environment and Development*
UNECE: *United Nations Economic Commission for Europe*
USDA: *United States Department of Agriculture*

Bibliography

Abaza, H., R. Bisset and B. Sadler (2004), *Environmental Impact Assessment and Strategic Environmental Assessment: Towards an Integrated Approach (1st Edition),* United Nations Environment Programme, Geneva, Switzerland. http://www.unep.ch/etu/publications/textONUBr.pdf (accessed 22 March 2007).

Administrative Review Council (2006), 'The Scope of Judicial Review'. Report to the Attorney-General, Report No.47, Barton, ACT, Australia.

AGB McNair (1994), 'Survey of Attitudes towards Land Use Issues for Shoalwater Bay Military Training Area'. Research Report Number 13, Commission of Inquiry into Shoalwater Bay, Capricornia Coast, Queensland, Final Report, **3**, 301–27, Commonwealth of Australia Australian Government Publishing Service, Canberra, Australia.

Amos, M. and S. Churches (1998), 'Declarations', in J.L. Davies, N.C. Seddon and G. Masel (eds) *The Laws of Australia*, **2.6**, Remedies in Judicial Review, Chapter 5, 179–96, Law Book Company, Pyrmont, NSW, Australia.

Andrues, M. (1989), 'Proof of cancer causation in toxic tort waste litigation: the case of determinacy versus indeterminacy', *Southern California Law Review*, **61**, 2075.

Arkin, S. (2003), 'Bringing California's anti-SLAPP statute full circle: to commercial speech and back again', *Western State University Law Review*, **31**, 1.

Arnold, C.A. and L.A. Jewell (2002), *Beyond Litigation: Case Studies in Water Rights Disputes*, Environmental Law Institute, Washington D.C., USA.

Ashby, E. (1978), *Reconciling Man with the Environment*, Stanford University Press, Stanford, California, USA.

Asimow, M., A. Bonfield and R. Levin (1998), *State and Federal Administrative Law (2nd Edition)*, West Publishing Co., St. Paul, Minnesota, USA.

Astor, H. and C. Chinkin (1992), *Dispute Resolution in Australia*, Butterworths, North Ryde, NSW, Australia.

Australian Law Reform Commission (1994), 'Freedom of Information', Issues Paper 12, Sydney, NSW, Australia.

Barber, B. (1984), *Strong Democracy: Participatory Politics for a New Age*, University of California Press, Berkeley, USA.

Barkas, J. (2005), 'Nuking freedom of information and community right to know: how post-9/11 secrecy politics could make America less safe', *Environs Environmental Law and Policy Journal*, **28**, 199.

Barnes, P. (2001), 'Regulating safety in an unsafe world (risk reduction *for* and *within* communities)', *Journal of Hazardous Materials*, **86**, 25–37.

Bates, Gerry (2006), *Environmental Law in Australia (6th Edition)*, LexisNexis Butterworths, Chatswood, NSW, Australia.

Beer, T. and F. Ziolkowski (1995), 'Environmental Risk Assessment: An Australian Perspective', Report 102, Department of the Environment and Heritage, Canberra, ACT, Australia.

Beierle, T.C. (1999), 'Using social goals to evaluate public participation in environmental decision making', *Policy Studies Review*, **16**, 3.

Bell, J. (1988), 'Proving causation', *Trial* (October), 50.

Bell, Stuart and Donald McGillivray (2006), *Environmental Law (6th Edition)*, Oxford University Press Inc., New York, USA.

Beloff, M.J. (2005), 'How green is judicial review?', *Judicial Review*, **10**, 93–110.

Berkes, F. (1999), *Sacred Ecology. Traditional Ecological Knowledge and Resource Management*, Taylor & Francis, Philadelphia and London.

Berkes, F., J. Mathias, M. Kislalioglu and H. Fast (2001), 'The Canadian arctic and the *Oceans Act*; the development of participatory environmental research and management', *Ocean & Coastal Management*, **44**, 451–69.

Bingham, G. and L.V. Haygood (1986), 'Environmental dispute resolution: the first ten years', *The Arbitration Journal*, **41**, 3–14.

Black, B. (1988), 'A unified theory of scientific evidence', *Fordham Law Review*, **56**, 595–695.

Black, B. and D. Lilienfield (1984), 'Epidemiological proof in toxic tort litigation', *Fordham Law Review*, **52**, 732–81.

Bonine, John (1998), 'Broadening "standing to sue" for citizen enforcement', Proceedings of the Fifth International Conference on Environmental Compliance and Enforcement, Volume 2, 259–64, 16–20 November 1998, Monterey, California, USA.

Bordone, R., M. Moffitt and F. Sander (2005), 'The Next Thirty Years: Directions and Challenges in Dispute Resolution', in M. Moffitt and R. Bordone (eds), *The Handbook of Dispute Resolution*, 507–19, Jossey-Bass, San Francisco, California, USA.

Boulle, Laurence (2005), *Mediation. Principles Process Practice (2nd Edition)*, LexisNexis Butterworths, Chatswood, NSW, Australia.

Boyd, D. (2003), *Unnatural Law: Rethinking Canadian Environmental Law and Policy*, 152, UBC Press, Vancouver, Canada.

Brans, E.H. (2005), 'Liability for damages to public natural resources under the 2004 EC Environmental Liability Directive', *Environmental Law Review*, **7**, 92–109.

Brennan, T. and R. Carter (1985), 'Legal and scientific probability of causation of cancer and other environmental diseases in individuals', *Journal of Health Politics, Policy and Law*, **10**, 33–80.

Brown, T. (1992), 'Techniques of Identifying Environmental Risks for Contaminated Sites', in A. Chambers and J. Whitehead (eds), *Environmental Risk Analysis*, 35–55, Occasional Paper No. 16, University of Newcastle, NSW, Australia.

Brunnée (2004), 'The United States and international environmental law: living with an elephant', *European Journal of International Law*, **15**, 617–49.

Buccini, John (2004), *The Global Pursuit of the Sound Management of Chemicals*, Report of the World Bank, Washington D.C., USA.

Buckley, R. (1989), 'Precision in Environmental Impact Prediction', Resource and Environmental Studies, Report No. 2, Centre for Resource and Environmental Studies, Australian National University, Canberra, Australia.

Buckley, R. (1991), 'How accurate are environmental impact predictions?', *Ambio*, **20**, 161.

Burger, E.J. (1988), 'Scientific information on judicial and administrative decisions', *Social Science & Medicine*, **27**, 1031–41.

Burgman, M., S. Ferson and H. Akcakaya (1993), *Risk Assessment in Conservation Biology*, Chapman & Hall, London, UK.

Caldwell, J.L. (1986), 'Discretionary remedies in administrative law', *Otago Law Review*, **6**, 245–59.

Camacho, A.E. (2005), 'Mustering the missing voices: a collaborative model for fostering equality, community involvement and adaptive planning in land use decisions. Installment 2', *Stanford Environmental Law Journal*, **24**, 269–330.

Campbell, T. (2006), 'Interfaith Community Organisation v Honeywell International: The third circuit affirms the citizen suit use of the Resource Conservation and Recovery Act and orders the $400 million cleanup of a polluted site', *Tulane Environmental Law Journal*, **19**, 121.

Carpenter, S. and W. Kennedy (1985), 'Managing environmental conflict by applying common sense', *Negotiation Journal*, **1**, 149–61.

Carson, Rachel (1962), *Silent Spring*, Houghton Mifflin, Cambridge, Massachusetts, USA.

Christie, E. (1990), 'Sustainable resource use, scientific terminology and environmental legislation: issues in statutory interpretation', *Environmental and Planning Law Journal*, **7**, 262–70.

Christie, E. (1991), 'The role of law and science in the resolution of disputes over factual evidence', *Environmental and Planning Law Journal*, **8**, 200–210.

Christie, E. (1992a), 'Toxic tort disputes: distinctive characteristics require special preparation for trial', *Law Society of Queensland Journal*, **22**, 279–93.

Christie, E. (1992b), 'Toxic tort disputes: proof of causation and the courts', *Environmental and Planning Law Journal*, **9**, 302–19.

Christie, E. (1993a), 'Alternative dispute resolution and public land disputes', *Queensland Alternative Dispute Resolution Review*, **1**, 12–16.

Christie, E. (1993b), 'Natural Resource Use Evaluation: Information Needs and Processes of Analysis. The Commission's Processes and Procedures', Commission of Inquiry into Shoalwater Bay, Capricornia Coast, Queensland, Commonwealth of Australia, Report No. 1, Australian Government Publishing Service, Canberra, Australia.

Christie, E. (1993c), 'Ecologically sustainable development and environmental dispute resolution', *Australian Dispute Resolution Journal*, **4**, 257–72.

Christie, E. (1993d), 'The precautionary principle and environmental decision-making', *Queensland Planner*, **33**, 10–14.

Christie, E. (1993e), 'The eternal triangle — the Biodiversity Convention, endangered species legislation and the precautionary principle', *Environmental and Planning Law Journal*, **10**, 470–85.

Christie, E. (2005), 'Toxic mould litigation: a new sunrise industry for lawyers?', *Insurance Law Journal*, **16**, 281–96.

Clark, R. (2000), 'Making EIA Count in Decision-Making', in M. Partidario and R. Clark (eds), *Perspectives on Strategic Environmental Assessment*, 15, CRC–Lewis, Boca Raton, Florida, USA.

Comly French, B. (1993), 'More effective citizen participation in environmental decision-making', *University of Toledo Law Review*, **24**, 389–423.

Commission of Inquiry into Fraser Island and the Great Sandy Region (1991), State of Queensland, Supplementary Report, Goprint and Division of Information, Department of Lands, Brisbane, Australia.

Commission of Inquiry into Shoalwater Bay, Capricornia Coast, Queensland, Commonwealth of Australia (1994a), Final Report, **1**, Australian Government Publishing Service, Canberra, Australia.

Commission of Inquiry into Shoalwater Bay, Capricornia Coast, Queensland, Commonwealth of Australia (1994b), Interim Report, Australian Government Publishing Service, Canberra, Australia.

Coppel, P. (2005), 'Environmental information in the new regime', *Journal of Planning & Environment Law*, January, 12–33.

Cormick, G., N. Dale, P. Edmond, G. Sigurdson and B. Stuart (1996), *Building Consensus for a Sustainable Future: Putting Principles into Practice*, National Round Table on the Environment and the Economy, Ottawa, Ontario, Canada.

Costanza, R. and C. Perrings (1990), 'A flexible assurance bonding system for improved environmental management', *Ecological Economics*, **2**, 57.

Council of the European Union (2005), Brussels European Council 16 and 17 June, 2005: Presidency Conclusions. Report 10255/1/05 REV 1.

Cowell, S. and C. Williams (2006), 'Conservation through buyer-diversity: a key role for not-for-profit land-holdings organisations in Australia', *Ecological Management & Restoration*, **7**, 5–20.

Craig, R.K. (2005), *Environmental Law in Context. Cases and Materials*, Thomson West, St. Paul, Minnesota, USA.

Cygan, A. (2003), 'Protecting the interests of civil society in community decision-making — the limits of Article 230 EC', *International and Comparative Law Quarterly*, **52**, 995–1012.

Dale, V., S. Brown, R Haeuber, N. Hobbs, N. Huntly, R. Naiman, W. Riebsame, M. Turner and T. Valone (2000), 'Ecological principles and guidelines for managing the use of land', *Ecological Applications*, **10**, 639–70.

Das, C.V. (2001), 'Judicial checks on government', *Commonwealth Law Bulletin,* **27**, 583–90.

Davies, J. Clarence (2000), 'Environmental ADR and public participation', *Valparaiso Law Review*, **34**, 389–401.

Defenders of Wildlife and Center for Wildlife Law (2000), 'The public in action; using state citizen suits statutes to protect biodiversity', Monograph prepared by the Defenders for Wildlife and Center for Wildlife Law.
http://www.defenders.org/states/publications/publicinaction.pdf (accessed 5 September 2006).

Deloria, Vine Jr (1996), 'Reserving to themselves: treaties and the powers of Indian tribes', *Arizona Law Review*, **38**, 963.

Denning, Lord (1979), *The Discipline of Law,* Butterworths, London, UK.

Department of the Environment and Heritage (2005), 'Annual Report 2004–2005. Biodiversity and Wildlife', Canberra, Australia.
http://www.deh.gov.au/about/publications/annual-report/04-05/outcome-1-biodiversity (accessed 31 May 2006).

Department of the Environment and Heritage (2006a), 'Wind Farm Report Warns of Risk to Orange-bellied Parrot', Media Release, 5 April 2006, by the Minister for the Environment and Heritage, Canberra, Australia. http://www.deh.gov.au/minister/env/2006/mr05apr206.html (accessed 18 April 2006).

Department of the Environment and Heritage (2006b), 'Varied Bald Hills Wind Farm Proposal Approved with Strict Conditions', Media Release, 21 December 2006, by the Minister for the Environment and Heritage, Canberra, Australia. http://www.deh.gov.au/minister/env/2006/mr21dec06.html (accessed 22 December 2006).

Department of the Environment and Heritage (2007), 'EPBC Act — Environment Assessment Process', Fact Sheet. http://www.deh.environment.gov.au/epbc-amenedments/index.htm/ (accessed 18 January 2007).

Department of the Environment and Water Resources (2007a), 'Public Participation in the EPBC Act', Canberra, Australia. http://www.environment.gov.au.epbc/public-involvement.html (accessed 19 July 2007).

Department of the Environment and Water Resources (2007b), 'Strategic Assessments and the EPBC Act', Canberra, Australia. http://www.environment.gov.au/epbc/assessmentsapprovals/strategic/index.html (accessed 11 August 2007).

Department of Families, Youth and Community Care (1996), 'Indigenous Land Interest Model: An Information Guide', Report prepared by the Social Impact Unit, Brisbane, Queensland, Australia.

Dernbach, John C. (2002a), 'Sustainable Development: Now More Than Ever', in John C. Dernbach (ed.), *Stumbling Towards Sustainability*, 45–61, Environmental Law Institute, Washington D.C., USA.

Dernbach, John C. (2002b), 'Synthesis', in John C. Dernbach (ed.), *Stumbling Towards Sustainability*, 1–42, Environmental Law Institute, Washington D.C., USA.

Dodson, Mick (1996), 'Power and cultural differences in native title mediation', *Aboriginal Law Bulletin*, **84**, 11.

Doll, R. and R. Peto (1981), *The Causes of Cancer*, Oxford University Press, New York, USA.

Doremus, H. (2005), 'Science plays defense: Natural resource management in the Bush administration', *Ecology Law Quarterly*, **32**, 249.

Doty, P. (2000), *Freedom of Information in the United States: Historical Foundations and Current Trends,* University of Texas, Austin, USA. www.utexas.edu/research/tipi/reports2/foia_doty.pdf (accessed 6 May 2007).

Douglas, Mary (1985), *Risk Acceptability According to the Social Sciences*, Russell Sage Foundation, New York, USA.

Downes, G. (2004), 'Concurrent expert evidence in the Administrative Appeals Tribunal', Paper presented to the Australasian Conference of Planning and Environment Courts and Tribunals, Hobart, 27 February 2004.

Duncombe, Sue and Judith Heap (2004), *Australasian Dispute Resolution*, Law Book Company, North Ryde, New South Wales, Australia.

Dyer, J., P. Fishburn, R. Steuer, J. Wallenius and S. Zionts (1992), 'Multiple criteria decision making, multiattribute utility theory: the next ten years', *Management Science*, **38**, 645–54.

Environmental Law Institute (2001), *Opportunities for Advancing Environmental Justice: An Analysis of U.S. EPA Statutory Authorities*, Environmental Law Institute, Washington D.C., USA.

Environmental Law Institute (2002), *A Citizen's Guide to using Federal Environmental Laws to Secure Environmental Justice*, Environmental Law Institute, Washington D.C., USA.

Environmental Protection Agency (2003), *Public Involvement Policy and Related Documents*, Washington D.C., USA. http://www.epa.gov/public involvement/public/index.htm (accessed 25 July 2006)

EU–US Biotechnology Consultative Forum (2000), Final Report *On the regulatory process and potential uses of biotechnology and general strategies for attaining global sustainable agriculture*. www.agbios.com/docroot/articles/2001008-A.pdf (accessed 14 May, 2007).

Evans, A.S. (1976), 'Causation and disease: the Henle–Koch–Evans Postulates revisited', *Yale Journal of Law and Medicine*, **49**, 179–95.

Fairweather, P. (1989), 'Environmental impact assessment: where is the science in EIA', *Search*, **20**, 141.

Farber, D. (1987), 'Toxic causation', *Minnesota Law Review*, **71**, 1219.

Ferrey, Steven (2004), *Environmental Law. Examples & Explanations (3rd Edition)*, Aspen Publishers, New York, USA.

Fiorino, D. (1990), 'Citizen participation and environmental risk: a survey of institutional mechanisms', *Science, Technology & Human Values*, **15**, 226–43.

Fisher, Roger, William Ury and Bruce Patton (1991), *Getting to Yes: Negotiating an Agreement Without Giving In (2nd Edition)*, Century Business, London, UK.

Fisher, Ronald (1926), 'The arrangement of field experiments', *J. Ministry Agric. Great Britain*, **23**, 503–13.

Fitzgerald, G.E. (2006), 'Down with adversarial behaviour', *Lawyers Weekly* Issue 280, 10 March.

Formby, J. (1987), 'EIA and the Tasmanian wood.chip controversy', *Environmental and Planning Law Journal*, **4**, 191.

France, T. (1990), 'NEPA — the next twenty years', *Land and Water Review*, **25**, 132.

Frickey, P.H. (1997), 'Adjudicating and the discontents: coherence and conciliation in Federal Indian law', *Harvard Law Review*, **110**, 1754.

Futrell, J. William (2002), 'Foreword', in John C. Dernbach (ed.), *Stumbling Towards Sustainability*, Environmental Law Institute, Washington D.C., USA.

Gaynor, F. (1950), MAX PLANCK, *Scientific Autobiography and Other Papers*, 33–4. 'Respectfully Quoted: A Dictionary of Quotations, 1989'. www.bartleby.com/73/1665.html (accessed 2 October 2006).

Gellhorn, E. and R. Levin (1997), *Administrative Law and Process in a Nutshell (4th Edition)*, West Publishing Co., St, Paul, Minnesota, USA.

Gerrard, M. (1994), *Whose Backyard, Whose Risk: Fear and Fairness in Toxic and Nuclear Waste Siting*, MIT Press, Cambridge, Massachusetts, USA.

Gilpin, A. (1986), *Environmental Planning. A Condensed Encyclopaedia,* Noyes Publications, Park Ridge, New Jersey, USA.

Gilpin, A. (1995), *Environmental Impact Assessment — Cutting Edge for the Twenty-First Century*, Cambridge University Press, Cambridge, UK.

Ginzburg, H. (1986), 'Use and misuse of epidemiological data in the courtroom: dealing with the limits of inferential and particularistic evidence in mass tort litigation', *American Journal of Law and Medicine*, **12**, 423–39.

Ginzburg L., L. Slobodkin, K. Johnson and A. Bindman (1982), 'Quasiextinction probabilities as a measure of input on population growth', *Risk Analysis*, **21**, 171–81.

Gold, S. (1986), 'Causation in toxic torts: burdens of proof, standards of persuasion and statistical evidence', *Yale Law Journal*, **96**, 376–402.

Goodman, E. (2000), 'Protecting habitat for off-reservation tribal hunting and fishing rights: tribal co management as a reserved right', *Environmental Law*, **30**, 279.

Gorczynski, D. (1991), *Insider's Guide to Environmental Negotiation*, Lewis Publishers, Inc., Chelsea, Michigan, USA.

Gough, S., W. Scott and A. Stables (2000), 'Beyond O'Riordan: balancing anthropocentrism and ecocentrism', *International Research in Geographical and Environmental Education*, **9**, 36–47.

GovTrack.us (2007), 'Threatened and Endangered Species Recovery Act of 2005'.
http://www.govtrack.us/congress/bill.xpd?bill=h109-3824 (accessed 6 July 2007).

Gusman, S. and P. Harter (1986), 'Mediating solutions to environmental risks', *Annual Review of Public Health*, **7**, 293–312.

Halpern, S. (1992), 'The United Nations Conference on Environment and Development: Process and Documentation', Academic Council for the United Nations System. http://www.ciesin.org/docs/008-585/unced-intro.html (accessed 29 September 2006).

Harding, A.J. (1989), *Public* Duties *and Public Law*, Clarendon Press, Oxford, UK.

Hardy, S. (1992), 'The Endangered Species Act: on a collision course with human needs', *Public Land Law Review*, **13**, 87.

Hare, B. (1991), 'Ecologically sustainable development', *Habitat*, **19**, 10.

Harter, P. (1982), 'Negotiating regulations: a cure for malaise', *Georgetown Law Journal*, **71**, 1–118.

Haskell, B., B. Norton and R. Constanza (1992), 'What is ecosystem health and why should we worry about it?', in R. Costanza, B. Norton and B. Haskell (eds), *Ecosystem Health. New Goals for Environmental Management*, Island Press, Washington D.C., USA.

Hayden, W. (1991), 'Opening address', National Conference on Public Issue Dispute Resolution, Commission of Inquiry into Fraser Island and the Great Sandy Region, Brisbane, Australia, February.

Herman, J. and I. Ryan (2004), 'The urgent need for reform of freedom of information in Australia', **114**, *FOI Review*, 62–6.

Hill, A.B. (1965), 'The environment and disease: association or causation?', *Proceedings of the Royal Society of Medicine*, **58**, 295–300.

Hilson, C. (2000), '*Laidlaw:* some observations from the UK/EU', *Journal of Environmental Law*, **12**, 381–4.

Hilson, C. (2005), 'Beyond rationality? Judicial review and public concern in the EU and the WTO', *Northern Ireland Legal Quarterly*, **56**, 320–41.

Hoffman, R. (1984), 'The use of epidemiological data in the courts', *American Journal of Epidemiology*, **120**, 190–202.

Holst, J. (1992), 'The unforeseeability factor: federal lands, managing for uncertainty, and the preservation of biological diversity', *Public Land Law Review*, **13**, 99.

Horn, L. (2005), 'Mediation of environmental conflicts', *Environmental and Planning Law Journal*, **22**, 369 –84.

House of Representatives Standing Committee on Environment and Heritage (2001), 'Public Good Conservation: Our Challenge for the 21st Century', Interim Report, The Parliament of the Commonwealth of Australia, Canberra, Australia.

Institute for Participatory Planning (1981), *Citizen Participation Handbook for Public Officials and Others Serving the Public (4th Edition)*, Laramie, Wyoming, USA.

International Risk Governance Council (2005), *Basic Concepts of Risk Characterisation and Risk Governance*, Geneva, Switzerland. www.irgc.org (accessed 8 December 2006)

Isaacson, P. (1986), 'Pollutant regulation and public sensibility', *Environmental Impact Assessment Review*, **6**, 229–32.

IUCN (The World Conservation Union) (2006), *2006 IUCN Red List of Threatened Species: Facts about Threatened Species*, IUCN, Gland, Switzerland. http://www.iucn.org/themes/ssc/redlist2006/threatened_species_facts.htm (accessed 8 May 2006).

Jasanoff, S. (1995), 'Product, Process, or Programme: Three Cultures and the Regulation of Biotechnology', in M. Bauer (ed.), *Resistance to New Technology,* 311–31, Cambridge University Press, Cambridge, UK.

Jones, M.A. (2006), 'Breach of Statutory Duty', in A.M. Dugdale and M.A. Jones (General Editors), *Clerk and Lindsell on Torts (14th Edition)*, 519–61, Sweet & Maxwell, London, UK.

Keeney, R. and H. Raiffa (1976), *Decisions with Multiple Objectives: Preferences and Value Tradeoffs*, John Wiley & Sons, New York, USA.

Kelly, C. (1995), 'Collateral abuse of process, malicious prosecution and civil SLAPP suits', *Impact*, **38**, 13–15.

Kelty, W. (1991), Paper on public interest environmental conflicts presented to the National Conference on Public Issue Dispute Resolution, Commission of Inquiry into Fraser Island and the Great Sandy Region, Brisbane, Australia, February.

Kermath, B. (2006), 'What is Sustainability?', World Conservation Union (IUCN) and Californian Institute of Public Affairs. http://www.uwsp.edu/CNR/gem/ambassador/What_is_sustainability.htm (accessed 8 May 2006).

Kirby, M. (1992), 'Mediation: current controversies and future directions', *Australian Dispute Resolution Journal*, **3**, 139–48.

Kiss, Alexandre and Dinah Shelton (2004), *International Environmental Law (3rd Edition)*, Transnational Publishers, Inc., Ardsley, New York, USA.

Kitzhaber, John A. (2004), 'Creating a society to match our scenery: resolving natural disputes in the 21st century', *Public Lands & Resources Law Review*, **25**, 1.

Klaarsem, C., M. Amdur and J. Doull (1986), *Caserett and Doull's Toxicology. The Basic Science of Poisons*, Macmillan, New York, USA.

Krimsky, S. (1993), 'Risk analysis and public policy', *Environment*, **35**, 5.

Kubasek, Nancy and Gary Silverman (2005), *Environmental Law (5th Edition)*, Pearson Prentice Hall, Upper Saddle River, New Jersey, USA.

Lahdelma, R., P. Salminen and J. Hokkanen (2000), 'Using multicriteria methods in environmental planning and management', *Environmental Management*, **26**, 595–605.

Lawrence, J. and D. Taylor (1990), 'Statistics and the precautionary principle', *Marine Pollution Bulletin*, **21**, 402.

Lawson, C. and R. Hindmarsh (2006), 'Releasing genetically modified canola into the environment — deconstructing a decision of the Gene Technology Regulator under the Gene Technology Act 2000 (Cth)', *Environmental and Planning Law Journal*, **23**, 22–59.

Leopold, Aldo (1949), *A Sand County Almanac*, Oxford University Press, London, UK.

Levidow, L. and S. Carr (2000), 'Normalizing novelty: regulating biotechnology risk in the U.S. EPA', *Risk: Health, Safety & Environment*, **11**, 9–34.

Levine, D.L. (2005), 'The future of defamation law. Courts & the media', Supreme Court of New South Wales, Sydney, Australia. http://www.lawlink.nsw.gov.au/lawlink/supreme_court/11_sc.nsf/pages/SCO_speech_1 (accessed 25 July, 2007).

Lippmann, M. and G. Thurston (1988), 'Exposure assessment: input into risk assessment', *Archives of Environmental Health*, **43**, 116.

McClenaghan, T., K. Cooper, L. Vanderlinden, P. Muldoon, A. Abelsohn, K. Khatter and K. Keenan (2002), 'Environmental standard setting and children's health in Canada: injecting precaution into risk assessment', *Journal of Environmental Law and Practice*, **12**, 245–79.

McDonald, J. and C. Jones (2003), *The Law of Freedom of Information*, Oxford University Press, Oxford, UK.

Mandel, G. (2006), 'The future of biotechnology litigation and adjudication', *Pace Environmental Law Review*, **23**, 83.

Manson, C. (2004), 'Zero-sum habitat game causes disputes', *Environmental Forum*, **21**, 49–50.

Marsden, S. (2002), 'Strategic Environmental Assessment: An International Overview', in S. Marsden and S. Dovers (eds), *Strategic Environmental Assessment in Australasia*, 1–23, The Federation Press, Leichhardt, NSW, Australia.

Martin, J. (1977), 'The proposed science court', *Michigan Law Review*, **75**, 1063.

Marton-Lefèvre, Julia (2007), *Release of the 2007 IUCN Red List of Threatened Species*, International Union for Conservation of Nature and Natural Resources, Gland, Switzerland.

http://www.iucn.org/en/news/archive/2007/09/12_pr_red_list_htm (accessed 13 September 2007).

Mattix, Carla and Kathleen Becker (2002), 'Scientific uncertainty under the National Environmental Policy Act', *Administrative Law Review*, **54**, 1125–65.

May, J.R. (2004), 'The availability of State environmental citizen suits', *Natural Resources & Environment*, Spring, 53–6.

Mayer, B. (2000), *The Dynamics of Conflict Resolution: A Practitioner's Guide*, Jossey-Bass, San Francisco, USA.

Melling, T. (1995), 'Bruce Babbit's use of governmental dispute resolution: a mid-term report card', *Land & Water Law Review*, **30**, 57.

Miller, J. (2000), 'The standing of citizens to enforce against violations of environmental statutes in the United States', *Journal of Environmental Law*, **12**, 370–81.

Moore, C. (1986), *The Mediation Process: Practical Strategies for Resolving Conflict*, Jossey-Bass, San Francisco, USA.

Moore, J. (1987), 'Not So Silent Spring', in G. Marco, R. Hollingsworth and W. Durham (eds), *Silent Spring Revisited*, 15–24, American Chemical Society, Washington D.C., USA.

Morgan, B. and G. Airo-Farulla (1997), 'No Evidence', in J.L. Davies, N.C. Seddon and G. Masel (eds), *The Laws of Australia*, **2.4**, Judicial Review of Administrative Action: General Grounds, Chapter 10, 145–52, Law Book Company, Pyrmont, NSW, Australia.

Mosley, G. (1988), 'Kakadu and Arnhem Land', in P. Figgis and G. Mosley (eds), *Australia's Wilderness Heritage*, 24–45, Lansdowne Publishing Pty, Ltd, Sydney, Australia and Australian Conservation Foundation.

National Biological Information Infrastructure (2007), 'United States regulatory agencies unified biotechnology website: frequently asked questions', Center for Biological Informatics, U.S. Geological Survey. http://wwwusbiotechreg.nbii.gov (accessed 1 May, 2007).

Nolan-Haley, N.M. (1992), *Alternative Dispute Resolution*, West Publishing Co., St. Paul, Minnesota, USA.

O'Faircheallaigh, C. (1999), 'Making social impact assessment count: a negotiation-based approach for indigenous peoples', *Society & Natural Resources*, **12**, 63–80.

Office of Technology Assessment Task Force (1988), *Identifying and Regulating Carcinogens*, Lewis Publishers, Inc., Chelsea, Michigan, USA.

Olson, Todd (1996), 'Biodiversity and Private Property; Conflict or Opportunity', in William Snape III (ed.), *Biodiversity and the Law*, 67–79, Island Press, Washington D.C., USA.

Olson, Todd, Dennis Murphy and Robert Thornton (1993), 'The Habitat Transaction Method: A Proposal for Creating Tradeable Credits in Endangered Species and Habitats', in H. Fischer and W. Hudson (eds), *Building Economic Incentives into the Endangered Species Act*, 27–36, Defenders of Wildlife, Washington D.C., USA.

O'Riordan, T. (1989a), 'The Challenge for Environmentalism', in R. Peet and N. Thrift (eds), *New Models in Geography*, 77–102, Unwin Hyman, London, UK.

O'Riordan, T. (1989b), 'Preface', in Jennifer Brown (ed.), *Environmental Threats: Perception, Analysis and Management*, vii, Belhaven Press, London, UK.

Otway, H. and D. Sims (1987), 'Criteria for technological choice', *Public Administration*, **65**, 131–43.

Otway, H. and D. von Winterfeldt (1982), 'Beyond acceptable risk: on the social acceptability of technologies', *Policy Science*, **14**, 247–56.

Paddle, G. (1990), 'Epidemiology', in D. Anderson and D. Connings (eds), *Experimental Toxicology. The Basic issues (1st Edition)*, 442, The Royal Society of Chemistry, Cambridge, UK.

Pandya, J. (2004), 'Biotechnology: a powerful revolution', *Risk Management Matters*, **2**, 1–3.

Parkowski, F.M. and M. Teichman (2004), 'Public participation in the Coastal Zone Act permitting process: too much of a good thing?', Delaware *Lawyer*, **22**, 18.

Passioura, J. (1973), 'Sense and nonsense in crop simulation', *Journal of the Australian Institute of Agricultural Science*, **39**, 181–3.

Paterson, Moira (2005), *Freedom of Information and Privacy in Australia*, LexisNexis Butterworths, Sydney, Australia.

Patterson, Lee Ann and Tim Josling (2001), *Regulating Biotechnology: Comparing EU and US Approaches*. www.ucis.pitt.edu/euce/pub/policypapers/2002-TransatlanticBiotech.pdf (accessed 14 May 2007).

Patton, Bruce (2005), 'Negotiation', in M. Moffitt and R. Bordone (eds), *The Handbook of Dispute Resolution*, 279–303, Jossey-Bass, San Francisco, USA.

Percival, R. (2006), 'Who's afraid of the precautionary principle?', *Pace Environmental Law Journal*, **23**, 21–81.

Perhac, R. (1998), 'Comparative risk assessment: where does the public fit in?', *Science, Technology & Human Values*, **23**, 221–41.

Pierce, R.J. Jr (2002), *Administrative Law Treatise (4th Edition)*, Volume 1, Aspen Publishers, New York, USA.

Pincus, C.W. (1988), 'Mixture of methods better to solve complex issues', *Australian Law News*, **23**, 19.

Plater, Z.J., R. Abrams and W. Goldfarb (2004), *Environmental Law and Policy: Nature Law and Society (3rd Edition)*, West Publishing Co., St. Paul, Minnesota, USA.

Power, M. and L. McCarty (1997), 'Fallacies in ecological risk assessment practices', *Environmental Science & Technology*, **31**, 370–75.

Pring, George and Penelope Canan (1988), 'Strategic lawsuits against public participation', *Social Problems*, **35**, 506.

Pring, George and Penelope Canan (1992), 'Strategic lawsuits against public participation ("SLAPPS"): an introduction for bench, bar and bystanders', *Bridgeport Law Review*, **12**, 937.

Pritzker, D.M. (1995), 'Regulation by consensus: negotiated rulemaking in the United States', *Commercial Dispute Resolution*, **1**, 217.

Quandt, J.E., G.M. Bozzer and J.K. Haynes (2005), 'The shifting sands of "citizen suits" after Cleveland Cliffs', *Michigan Bar Journal*, November, 32–5.

Rall, D. (1988), 'Laboratory animal toxicity and carcinogenesis: testing, underlying concepts, advantages and constraints', *Annals New York Academy of Science*, **534**, 78–83.

Ramirez, Jose and Ivan Galindo (2004), 'Risk assessment. A perspective from the south', *Risk Management Matters*, **2**, 4–6.

Red Lodge Clearing House (2007), 'Endangered Species Act (ESA)', Department of the Interior, Washington D.C., USA. http://www.redlodgeclearinghouse.org/legislation/esa7.html (accessed 6 July 2007).

Redgwell, Catherine (2006), 'Biotechnology, Biodiversity and Sustainable Development: Conflict or Congruence?', in F. Francioni and T. Scovazzi (eds), *Biotechnology and International Law*, 61–79, Hart Publishing, Portland, Oregon, USA.

Resource Assessment Commission (1993), 'Coastal Zone Inquiry', Final Report, Australian Government Publishing Service, Canberra, Australia.

Riley, T. (1987), 'Accountability of government: an international perspective', *FOI Review*, **11**, 54.

Rogus, Lee Ann (1996), 'The Basel Convention and the United States', *New England International and Comparative Annual*, **2**, 18pp. http://www.nesl.edu/intljournal/vol2/basel.htm (accessed 12 March 2007).

Royal, C. Ashley (1987), 'The defence of medical causation', *Trial* (October), 44.

Royal Commission into the Use and Effects of Chemical Agents on Australian Personnel in Vietnam (1985), Report of Justice Evatt, Commonwealth of Australia, Canberra.

Rutherford, Susan and Karen Campbell (2004), 'Time well spent? A survey of public participation in Federal environmental assessment panels', *Journal of Environmental Law and Practice*, **15**, 71–83.

Sackville, R. (2000), 'The limits of judicial review of executive action – some comparisons between Australia and the United States', *Federal Law Review*, **28**, 315–30.

Sample, V.A. (1991), 'Assessing cumulative environmental impacts: the case of national forest planning', *Environmental Law*, **21**, 839–62.

Sampson, T. (2003), 'Environmental risk assessment of GMOs under Directive 2001/18: an effective safety-net or a "collective illusion" ', *European Intellectual Property Review*, Issue 2, 79–90.

Saunders, J. Owen (1990), *The Legal Challenge of Sustainable Development*, Canadian Institute of Resources Law, Calgary, Alberta, Canada.

Schindler, W. (1976), 'The impact a statement boondoggle', *Science*, **192**, 509.

Schmeuli, Deborah (2003), 'Conflict Assessment', in G. Burgess and H. Burgess (eds), *Beyond Intractability*, Conflict Research Consortium, University of Colorado, Boulder. Posted October 2003.
http://www.beyondintractability.org/essay/conflict_assessment (accessed 22 June 2007).

Schoombee, H. (1998), 'Prerogative Orders', in J.L. Davies, N.C. Seddon and G. Masel (eds), *The Laws of Australia*, **2.6**, Remedies in Judicial Review, Chapter 4, 161–78, Law Book Company, Pyrmont, NSW, Australia.

Schuck, P.H. (1987), *Agent Orange on Trial: Mass Toxic Disasters in the Court*, Belknap Press, Cambridge, Massachusetts, USA.

Schwartzbauer, E. and S. Shindell (1988), 'Cancer and the adjudicative process: the interface of environmental protection and toxic tort law', *American Journal of Law and Medicine*, **14**, 1–67.

Scott, Joanne (2004), 'European regulation of GMOs: thinking about judicial review in the WTO', *Current Legal Problems*, **57**, 117–47.

Secretariat of the Convention on Biological Diversity (2007a), 'Notes on Ratification, Accession, Acceptance and Approval', Montreal, Canada.
http://www.cbd.int/doc/ref/ratification.htm (accessed 3 August 2007).

Secretariat of the Convention on Biological Diversity (2007b), 'List of Parties to the Convention on Biological Diversity and the Cartagena Protocol on Biosafety', Montreal, Canada.
http://www.cbd.int/world/parties.asp (accessed 3 August 2007).

Sedley, Lord Justice (2004), 'The last 10 years' development of English public law', *Australian Journal of Administrative Law*, **12**, 9–21.

Selby, H. (1992), 'Lawyers should support a public-interest law centre', *Australian Law News*, **27**, 22–5.

Selinger, B. (1984), 'Science in the witness box', *Legal Services Bulletin*, **9**, 108.

Severn, D. (1987), 'Exposure assessment', *Environmental Science and Technology*, **21**, 1160.

Shaffer, Mark (1987), 'Minimum Viable Populations: Coping with Uncertainty', in M.E. Soulé (ed.), *Viable Populations for Conservation*, 69–86, Cambridge University Press, Cambridge, UK.

Sheller, J. (1999), 'Judgement writing', *The Judicial Review*, **4**, 127.

Shere, M. (1995), 'The myth of meaningful environmental risk assessment', *Harvard Environmental Law Review*, **19**, 409–92.

Sinclair, A.J. and M. Doelle (2003), 'Using law as a tool to ensure meaningful public participation in environmental assessment', *Journal of Environmental Law and Practice*, **12**, 27–53.

Slovic, P. (1993), 'Perceived risk, trust and democracy', *Risk Analysis*, **13**, 675–82.

Smith, E.T. (1987), 'Environmental Conflict Management', in D. Sandole and I. Sandole-Staroste (eds), *Conflict Management and Problem Solving: Interpersonal to International Applications*, 149–57, New York University Press, New York, USA.

Snape, William III (1996), 'Biodiversity's Safety Net: Saving Endangered Species', in William Snape III (ed.), *Biodiversity and the Law*, 33–42, Island Press, Washington D.C., USA.

Sourdin, Tania (2005), *Alternative Dispute Resolution (2nd Edition)*, Lawbook Co., Pyrmont, NSW, Australia.

Stanton, K., P. Skidmore, M. Harris and J. Wright (2003), *Statutory Torts*, Sweet & Maxwell, London, UK.

Steiner, A. (2006), *Release of the 2006 IUCN Red list of Threatened Species*, International Union for Conservation of Nature and Natural Resources, Gland, Switzerland. http://www.iucn.org/en/news/archive/2006/05/02_pr_red_list_en.htm(accessed 2 May 2006).

Stewart, D. and J. McColl (1994), 'The Resource Assessment Commission: an inside assessment', *Journal of Environmental Management*, **1**, 29.

Stookes, Paul (2005), *A Practical Approach to Environmental Law*, Oxford University Press Inc., New York, USA.

Street, L. (2005), 'Foreword to the First Edition', in Tania Sourdin, *Alternative Dispute Resolution (2nd Edition)*, v–vi, Lawbook Co., Pyrmont, NSW, Australia.

Stuart, B. (1992), 'Land Claim Agreements as a Mechanism for Resolving Resource-Use Conflicts', in M. Ross and J. Owen Saunders (eds), *Growing Demands on a Shrinking Heritage: Managing Resource Use*

Conflicts, 130–54, Canadian Institute of Resources Law, Calgary, Alberta, Canada.

Susskind, Lawrence (1985), 'The siting puzzle. Balancing economic and environmental gains and losses', *Environmental Impact Assessment Review*, **5**, 157–63.

Susskind, Lawrence (2005), 'Consensus Building and ADR: Why They Are Not the Same Thing', in M. Moffitt and R. Bordone (eds), *The Handbook of Dispute Resolution*, 358–70, Jossey-Bass, San Francisco, USA.

Susskind, Lawrence and Merrick Hoben (2004), 'Making regional policy dialogues work: a credo for metro-scale consensus building', *Temple Environmental Law & Technology Journal*, **22**, 123–38.

Susskind, Lawrence and Jennifer Thomas-Larmer (2000), 'Conducting a Conflict Assessment', in L. Susskind, S. MacKearnan and J. Thomas-Larmer (eds), *The Consensus Building Handbook: A Comprehensive Guide to Reaching Agreement*, 99–135, Sage Publications, Thousand Oaks, California, USA.

Susskind, Lawrence and Alan Weinstein (1980), 'Towards a theory of environmental dispute resolution', *Boston College of Environmental Affairs Law Review*, **9**, 311–57.

Susskind, Lawrence, Paul Levy and Jennifer Thomas-Larmer (2000), *Negotiating Environmental Agreements*, Island Press, Washington D.C., USA.

Tarlock, Dan (2002), 'Biodiversity and Endangered Species', in John C. Dernbach (ed.), *Stumbling Towards Sustainability*, 311–26, Environmental Law Institute, Washington D.C., USA.

The Nobel Foundation (2006), 'Medicine 1948', Stockholm, Sweden. http://www.nobelprize.org/nobel_prizes/medicine/laureates/1948/ (accessed 2 October 2006).

Thornton, Scott (1991), 'Cumulative impacts in environmental review: the New York standpoint', *Pace Environmental Law Review*, **9**, 253–77.

Thornton, Justine and Silas Beckwith (2004), *Environmental Law*, Sweet & Maxwell, London, UK.

Tilleman, W. (1995), 'Public participation in the environmental impact assessment process: a comparative study of impact assessment in Canada, the United States and the European Community', *Columbia Journal of Transnational Law*, **33**, 337–439.

Tromans, S. (2006), 'Environmental judicial review', *Environmental Law*, **30**, 26–38.

Turner, John (2001), *Biotechnology Deskbook*, Environmental Law Institute, Washington D.C., USA.

Tynan, E. (2005), 'Toxic chemicals — a growing global threat', *Annual Review (July 2004–June 2005) Environmental Matters – The World Bank Group*, 30–31.

Union of Concerned Scientists (2006), 'Science in the Endangered Species Act is Threatened', Cambridge, Massachusetts, USA. http://www.ucsusa.org (accessed 31 May 2006).

United Nations Economic Commission for Europe (2006), 'Multilateral Treaties Deposited with the Secretary General **13**. Convention on Access to Information, Public Participation in Decision-Making and Access to Justice in Environmental Matters (Aarhus Convention)', Geneva, Switzerland. http://www.unece.org/env/pp/treaty.htm (accessed 15 November 2006).

United Nations Environment Programme (2005), 'Basel Convention Technical guidelines, Series, SBC, Nr 2005/1'. http://www.basel.int (accessed 3 October 2006).

United Nations General Assembly (2007), 'UN News Service, United Nations adopts Declaration on Rights of Indigenous Peoples', United Nations, New York. http://www.un.org/apps/news/printnews.asp?nid=23794 (accessed 17 September 2007).

United States National Academy of Sciences (1983), *Risk Assessment in the Federal Government: Managing the Process*, National Academy Press, Washington D.C., USA.

US Department of Energy (2007), 'DOE Environmental Policy & Guidance. Resource Conservation and Recovery Act', Office of Health, Safety and Security, Washington D.C., USA. hss.energy.gov/nuclearsafety/nsea/oepa/policy/rcra.html (accessed 12 March 2007).

US Environmental Protection Agency (2007), 'National Environmental Policy Act (NEPA): Basic Information', Washington D.C., USA. http://www.epa.gov/compliance/basics/nepa.html (accessed 19 March 2007).

US Fish and Wildlife Service (2004), 'Safe Harbor Agreements for Private Landowners, U.S. Fish and Wildlife Service Endangered Species Program 703/358-2105', Washington D.C., USA. http://endangered.fws.gov (accessed 2 June 2006).

Vilcheck, Michelle (1991), 'The controls on the transfrontier movement of hazardous waste from developed to developing countries: the goal of a level playing field', *Journal of International Law and Business*, **11**, 643.

Vogt, D. and M. Parish (2001), *Food Technology in the United States: Science, Regulation and the Issues*, Congressional Research Service Report for Congress, USA.

www.ncseonline.org/nle/crsreports/science/st-41.pdf (accessed 24 January 2007).

Wade, H.W.R. and C.F. Forsyth (2004), *Administrative Law (9th Edition)*, Oxford University Press, Oxford, UK.

Watfa, N., S. Awan and R. Goodson (1998), *Chemical Risk Assessment and Occupational Hygiene Preventive Measures in Small and Medium-Sized Enterprises*, International Labour Organisation, Geneva, Switzerland. www.ilo.org/public/english/protection/safework/papers/smechem/index.htm (accessed 3 October 2006).

Wagner, W.E. (1986), 'Trans-science in torts', *Yale Law Journal*, **96**, 428–49.

Wald, P.M. (1984), 'The Freedom of Information Act: a short case study in the perils and paybacks of legislating democratic values', *Emory Law Journal*, **33**, 649.

Walters, Brian (2003), *Slapping on the Writs,* University of New South Wales Press, Sydney, NSW, Australia.

Weinstein, J. (1986), 'Improving expert testimony', *University of Richmond Law Review*, **20**, 473.

Weston, J. (2002), 'From Poole to Fulham: a changing culture in UK environmental impact assessment decision making?', *Journal of Environmental Planning and Management*, **45**, 425–43.

White, J. (1980), 'Machiavelli and the bar. Ethical limitations on lying in negotiation', *American Bar Foundation Research Journal*, **5**, 926–38.

Wigmore, J.H. (1983), *Evidence in Trials at Common Law*, Volume 1, Chapter 2, Little, Brown and Company, Boston, Massachusetts, USA.

Wilcox, M. (1985), 'The role of environmental groups in litigation', *Adelaide Law Review*, **10**, 41–8.

Wolf, Susan and Anna White (1997), *Principles of Environmental Law (2nd Edition)*, Cavendish Publishing Ltd, London, UK.

World Commission on Environment and Development (1987), *Our Common Future*, Oxford University Press, Oxford, UK.

World Health Organisation (2007), *Toxic Hazards. Agrochemicals, Health and Environment*, Geneva, Switzerland. http://www.who.int/heli/risks/toxics/chemicals/en/print.html (accessed 13 February 2007).

Wynne, B. (1989), 'Building Public Concerns into Risk Management', in J. Brown (ed.), *Environmental Threats: Perceptions, Analysis and Management*, 118–33, Belhaven Press, London, UK.

Table of authorities — statutes, regulations, conventions

Australia: Statutes and Regulations

(a) Federal Legislation

(b) Federal Regulations

(c) Legislation: States and Territories

(d) Regulations: States and Territories

Canada: National Legislation

The United Kingdom: Statutes and Regulations

(a) National Legislation

(b) National Regulations

European Union Law

(a) Treaties

(b) Regulations

(c) Directives

[6] Followed by the "Maastricht Treaty on European Union" (1992), then modified by the "Treaty of Amsterdam" (1997); the EC and EU treaties were consolidated in 2002 (Kiss and Shelton, 2004). In June 2007, agreement was reached for a new Reform Treaty —"Treaty on the Functioning of the Union".

United States of America: Statutes and Regulations

(a) Federal Legislation

(b) Federal Bills

International Conventions and Other Instruments

Table of cases

New Zealand

United Kingdom

European Court of Justice

USA

Index